THE OPEN UNIVERSITY

Science:
A Second Level Course

S299
GENETICS

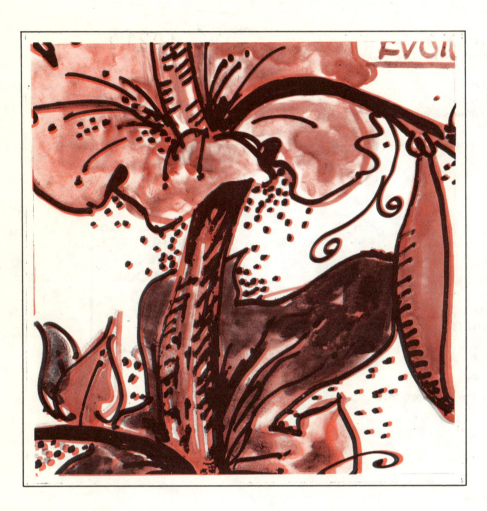

Prepared by a Course Team for the Open University

THE OPEN UNIVERSITY PRESS

Course Team

Chairman and General Editor
Steven Rose

Unit Authors
Norman Cohen (*The Open University*)
Terence Crawford-Sidebotham (*University of York*)*
Denis Gartside (*University of Hull*)
David Jones (*University of Hull*)
Steven Rose (*The Open University*)
Derek Smith (*University of Birmingham*)
Mike Tribe (*University of Sussex*)
Robert Whittle (*University of Sussex*)

*Consultant

Editor
Jacqueline Stewart

Other Members
Bob Cordell (*Staff Tutor*)
Jean Holley (*Technician*)
Stephen Hurry
Roger Jones (*BBC*)
Aileen Llewellyn (*Course Assistant*)
Michael MacDonald-Ross (*IET*)
Jean Nunn (*BBC*)
Pat O'Callaghan (*Evaluation*)
Jim Stevenson (*BBC*)

The development of this Course was supported by a grant from the Nuffield Foundation.

The Open University Press,
Walton Hall, Milton Keynes.

First published 1976.

Designed by the Media Development Group of the Open University.

Set by Composition House Ltd, Salisbury, Wiltshire.

Printed in Great Britain by Eyre and Spottiswoode Limited, at Grosvenor Press, Portsmouth.

ISBN 0 335 04280 5

This text forms part of an Open University Course. The complete list of Units in the Course appears at the end of this text.

For general availability of supporting material referred to in this text please write to the Director of Marketing, The Open University, P.O. Box 81, Walton Hall, Milton Keynes, MK7 6AT.

Further information on Open University Courses may be obtained from the Admissions Office, The Open University, P.O. Box 48, Walton Hall, Milton Keynes, MK7 6AB.

1 What is Genetics?

Contents

List of scientific terms used in Unit 1

Introduced in S100*	Developed in this Unit†	Page No.
amino acid	allele	19
blood group	chromosomes	24
diploid	classical (formal) genetics	7
DNA strand	developmental genetics	26
fertilization	dominant	16
gamete	eugenics	30
gene	first and second filial generations	14
genetic code	genetic counselling	30
genetic information	genetic programme for development	27
genotype	heterozygous advantage	25
haploid	interaction between genotype and	
heterozygous	environ	27
homologous	molecular genetics	7
homozygous	partial dominance	21
mutation	pedigree analysis	20
natural selection	population genetics	7
nucleotide base	progeny testing	11
ovum	recessive	16
phage	reciprocal mating	11
phenotype	selection pressure	26
pollen	self	13
protein chain		
sickle-cell anaemia		
zygote		

* The Open University (1971) S100 *Science: A Foundation Course*, The Open University Press.

† Many of the terms used in this Unit will be further developed in the rest of the Course.

Objectives for Unit 1

After studying this Unit you should be able to:

1   Define, recognize the best definition of, and place in the correct context, the items in the list of scientific terms above.
(ITQs 1–3)

2   Given a brief description of a genetic phenomenon, describe in outline, or select from a given list, appropriate methods for studying various aspects of the phenomenon.
(ITQ 8)

3   Given suitable data, test whether inheritance is of classical Mendelian type or not.
(SAQ 3)

4   Interpret or construct lineage charts using a simple notation.
(ITQs 4–6; SAQs 1 and 2)

5   Interpret pedigrees in terms of simple Mendelian genetics.
(ITQs 3, 7 and 9; SAQ 2)

6   Cite three examples of an interaction between genetics and society.

Study guide for Unit 1

This Unit is basically intended to give you a broad overview of what genetics is about, and to introduce some of the terminology to you. This means that frequently the treatment of topics is superficial. Should you not understand everything, do not worry—all the topics will be dealt with more thoroughly in the rest of the Course. Reference to the Objectives will help you decide what you need to know from this Unit. In-text questions (ITQs) occur at intervals in the narrative. We strongly advise you to attempt them and immediately to check your answers against those given on pp. 49–51; if you do not, you may not be able to follow the rest of the text. Self-assessment questions (SAQs) are grouped together at the back of the text.

In addition to the text, the other components of this Unit should be studied carefully. The TV programme can be viewed before reading the text, but the radio programme will probably be better understood if you have studied at least up to and including Section 1.2. As you will probably find that you can achieve the Objectives of this Unit fairly quickly, you should be able to study Mendel's paper, 'Experiments on plant hybrids' (see Appendix 1), in association with this text, and maybe even begin work on the history and social relations text*. Note that a preliminary reading of Mendel's paper should make listening to Radio programme 1 more rewarding.

1.0 Introduction to Unit 1

A few years ago a German farmer was plaintiff in a bizarre court case. The farmer, a married man, had two grown-up sons; one, the elder, was a strong vigorous man, the other, the younger, was weak, unmanly—a total disappointment to the farmer, himself strong and virile, 'like his elder son'. The younger son could obviously not have been sired by the farmer and so the farmer asked the German courts to prove just that and justify his legally disinheriting the young weakling. The court duly ordered blood samples to be taken from the farmer, his wife and both sons, to analyse them for blood-group characteristics known to be inherited in a predictable way. The tests indeed showed dubious paternity of one son—the elder, strong one; only with a likelihood of about 1 in 1 000 could he be the farmer's son. On the other hand, there was a much stronger possibility that the young one was the farmer's son.

This anecdote illustrates several points relevant to this Course. The farmer's obsession about the paternity of the younger son stems from the common observation that 'like begets like'. The observation probably goes well back into prehistoric times and has been a major influence on many early theories of inheritance. Most of these early theories assume, at least implicitly, as did the farmer, that the male is the significant parent in inheritance, particularly of desirable characteristics. The courts, however, did not need to argue on the basis of such prejudices; they were able to apply more objective scientific criteria. That they were able to do so is a tribute to the study of inherited characteristics—*the science of genetics*—over the last 100 years. It is genetics—its development, its current and future use—that this Course is about.

Let us take a closer look at the contribution genetics made to the paternity case. It enabled conclusions to be reached with a well-defined probability, but not absolute certainty—nevertheless, an advance over baseless prejudice. The contribution depended on both knowledge of the existence of particular blood groups in the parents and the probability of their inheritance by the sons. So genetics has helped, in some instances, to clarify the rules of biological inheritance and to replace bigotry by biology. However, do not be misled into thinking that genetics, which after all represents the thinking of geneticists, has developed over the last 100 years free of prejudice. Throughout these 100 years, and indeed to this very day, just as genetics has influenced our thinking and everyday lives in areas such as animal breeding, crop production and medicine, and promises (or threatens) to influence also social

---

* The Open University (1976) S299 HIST *The History and Social Relations of Genetics*, The Open University Press. This text is to be studied in parallel with the Units of the Course. From now on we shall refer to it by its code, *HIST*.

welfare and education policies, so genetics itself has been influenced by social mores and prejudices. These interactions between genetics and society will become apparent during this Course.

Let us now take a more detailed look at what genetics as a 'pure' science involves and what its producers, the geneticists, are interested in.

1.1 Sickle-cell anaemia: approaches to genetics

It can be argued that there are as many topics of interest to geneticists as there are geneticists. Nevertheless, there are, broadly speaking, a few definable categories of genetic topics and similarly a few categories of geneticists; whether the topics begat the geneticists or the geneticists the topics is debatable! The Course as a whole will span right across these categories but, though a little arbitrary, it is now useful to indicate the boundaries of these different areas in genetics and we can best do this by reference to a particular topic. A topic that crosses many areas in genetics is *sickle-cell anaemia*.

As those of you who have studied S100 will recall, sickle-cell anaemia is a disease occurring commonly in people in West and Central Africa, and occasionally in certain Mediterranean, Near Eastern and Indian areas. The symptoms of the disease are that severely affected individuals are anaemic and suffer blockages of the circulatory system. This is because their red blood cells tend to be sickle-shaped, rather than the normal round (Fig. 1).

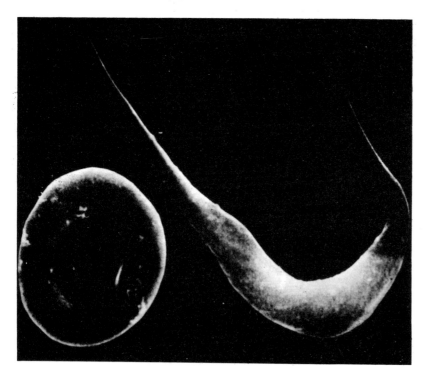

Figure 1 Cells shown in a scanning electron micrograph. Cell A is a normal red blood cell; Cell B is a sickle-cell.

Genetic studies have demonstrated that: *the disease is inherited in classical Mendelian fashion; the sickle-cell trait is due to a single recessive point mutation leading to the substitution of a single amino acid in the β-chain of haemoglobin; although the homozygote has a low chance of survival, the heterozygote has a selective advantage in areas with a high incidence of malaria, and when such pressures are absent the frequency of the sickle-cell allele in the population falls.* This summarizes what we now know about the genetic aspects of sickle-cell anaemia. By the end of this Course its full significance should have become apparent. It has taken a great deal of work over the 60 or so years since Herrick first noticed the sickle-shaped cells in some individuals, actually to establish the facts, and a lot more work to develop the concepts and techniques employed. Broadly, the concepts fall into three areas of genetics—classical (formal), molecular and population (evolutionary).

Now any one problem, such as that of sickle-cell anaemia, raises questions for all areas of genetics, and to answer it requires the use of the approaches and tools of

6

each. A geneticist studying sickle-cell anaemia these days is likely to find it necessary to understand all aspects of the problem and use many different tools. Indeed, it can be argued that the advances of recent years have fused together areas that in the past has been distinct. None the less, at least for much of the century or so since genetics began as a science, there have been these three discernable separate approaches (see *HIST*), and although we hope that by the end of this Course we shall have succeeded in unifying them to your satisfaction, we begin by asking which questions are seen as important by geneticists working in areas that have traditionally been regarded as classical, molecular and population genetics.

1.1.1 The classical geneticist

The concept 'like begets like' is probably one that goes well back into human history: we all expect dogs to give birth to pups and cats to kittens. Children also generally resemble their parents. Similarly, if one examines individual characteristics or traits, it is generally accepted that certain things 'run in families'—deafness, blue eyes, baldness, diabetes, good brains, money, etc. But to what degree are traits that are generally accepted as running in families really *biologically inherited*? This is essentially the first question the 'classical' (or 'formal') geneticist asks when studying a supposedly inherited characteristic. So the classical geneticist would, for example, ask about sickle-cell anaemia:

classical genetics

1   Do all members of a particular family inherit sickle-cell anaemia?

2   Is it only members of one sex that inherit the disease?

3   Does the trait appear in each generation of the family?

4   Is there any connection between the disease and the order of birth of the children?

1.1.2 The molecular geneticist

The molecular geneticist starts from the (not unreasonable) assumption that there must be some actual physical structures that are responsible for inheritance. That is, there must be molecules that convey genetic information (information specifying the characteristics of the organism) from one generation to the next. The molecular geneticist is, therefore, interested in four things:

molecular genetics

1   What is the nature of the molecules that carry genetic information?

2   How is this information transmitted from one generation to the next?

3   How is the information used to 'make' all the inherited characteristics of the organism?

4   Where changes occur in the genetic information, how do these come about?

So, in sickle-cell anaemia, he or she might well begin by asking:

1   What is the 'molecular reason' for the difference between the normal and the anaemic individual? In other words, what molecules are defective in the anaemic individual?

2   What is the nature of the difference in the molecules carrying genetic information between the normal and the anaemic individuals?

3   What 'agents' are responsible for the differences in the molecules carrying genetic information in the normal and anaemic individuals?

1.1.3 The population and evolutionary geneticist

The population geneticist is not primarily interested in the occurrence of a trait in an individual, nor the molecular mechanisms of occurrence or transmission of that trait as such. He or she is much more interested in the frequency or incidence of a particular trait in a population as a whole. In this context, *population* means a group of potentially interbreeding organisms isolated from other groups or populations of the same species by such factors as geographical or social barriers. The population geneticist's interest in sickle-cell anaemia would, therefore, be:

population genetics

1   What is the frequency of incidence of the disease among various populations?

2 Why, for example, does it occur in Africa and the Near East and not in Japan?

3 What are the genetic connections, if any, between populations that have the disease?

4 What factors have influenced or now influence the frequency of the disease among these populations?

These questions might also be asked by someone interested in the mechanisms of evolution, a so-called *evolutionary* geneticist. He or she is interested in the genetic changes that occur in populations, and thus provide the raw material on which evolution operates. To an evolutionary geneticist, sickle-cell anaemia could be taken as a model system of small-scale evolution.

Entry test for Sections 1.2–1.7

These Sections, particularly Sections 1.2.3, 1.2.4 and 1.4.2, require some prior knowledge, for example, from S100, of certain aspects of genetics. We therefore advise you to attempt this test, and then to check your answers against those given on p. 49 *before* continuing with this Unit.

Classify each of the following statements as either *true* or *false*:

1 The two strands of the DNA helix are held together by hydrogen bonds between the phosphate groups.

2 As a result of mitosis, the number of chromosomes per cell is reduced by half.

3 The alteration of a characteristic of an organism because of a change in its DNA can be termed a mutation.

4 The sequence of nucleotide bases in the DNA is determined by the sequence of amino acids in the polypeptide chain.

5 In a diploid organism, gamete cells contain twice as many chromosomes as other cells.

6 An individual that is homozygous for a gene contains two identical copies of that gene.

7 DNA replication occurs by way of a conservative system.

8 During meiosis, chromosomes are arranged homologously in pairs along the central equator of the cell.

9 A zygote arises through the mitotic division of a single gamete.

10 If a polypeptide chain is 100 amino acids in length, the mRNA corresponding to it is probably 200 nucleotides long.

Now check your answers against those given on p. 49.

1.2 The tools of classical genetics: 'It is a wise father that knows his own child'

1.2.1 Lineage studies

To put any suspicion that a particular trait 'runs in families' on a more objective footing, it is first necessary to examine the trait in as many individuals and in as many families as possible, thus tracing the inheritance of the trait in several families. Such studies are named *lineage studies*. So, for example, one can ascertain whether sickle-cell anaemia occurs in all members of a family or just a few; over how many generations it occurs; whether it occurs in just the members of one sex, and so on.

However, lineage studies in humans suffer from several obvious difficulties:

1 Diagnosis of the trait under investigation may not always be simple.

2 It is unusual for more than three generations to be alive at the same time.

3 Not all members of a family may be available for investigation.

4 As with the German farmer, supposed familial relationships may not always be true.

5 Few such studies have been carried out until recently and, as human generations are long, this means that the inheritance of only a few traits has been studied for more than three or four generations.

All this means that data collection often depends on family records or simple memory—neither very dependable. However, in certain aristocratic families unusual traits (or indeed whims) were noted carefully. The families could afford what doctors there were available; the family chroniclers recorded the information and sometimes from such things as family portraits, the traits are revealed, as in the case of the famous Habsburg lip, which has been recorded in this royal Austrian family over many years (Fig. 2).

Figure 2 The expression of the *Habsburg lip*, over four centuries. (a) Emperor Maximilian I (1449–1519); (b) Emperor Charles V (1500–1558), grandson of (a); (c) Archduke Charles (1771–1847); (d) Archduke Albrecht (1817–1895), son of (c).

In more recent years, it has been possible to construct complete *lineage charts* or *pedigrees* for several families carrying traits. From Figure 3 you can see that *albinism* (a lack of pigment in the skin, hair and eyes) 'runs in families'.

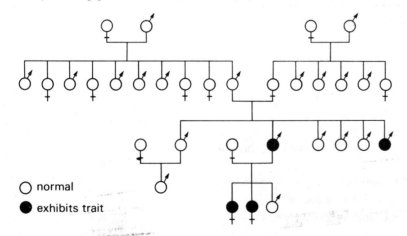

○ normal
● exhibits trait

Figure 3 A pedigree chart of a family group in which albinism has occurred.

The conventions adopted in this pedigree are worth examining, as we shall use them throughout the Course.

 is female; ♂ is male.

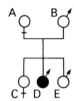 indicates a marriage (union), females normally being placed on the left.

indicates offspring; the order of birth is from left to right.

An open circle is a 'normal' person; a filled circle is a person exhibiting the trait under examination.

So, for example,

Figure 4

means that woman A married man B. They had three children. The eldest was a girl (C). The middle child, a boy (D), exhibited the trait.

Traits such as the Habsburg lip, albinism and sickle-cell anaemia initially arise as a result of *mutation*. That is, the individuals concerned (*mutants*) occur by some change from what is considered the normal, or *wild type*.

mutation

One can set about constructing pedigrees for sickle-cell anaemia fairly readily as the trait is quite common and so its inheritance can be followed in several families. Several such charts were constructed from as early as 1923. However, because of poor diagnosis (point 1 above), many such charts were incomplete. This is basically because two 'forms' of the trait occur in families and early studies did not even recognize the existence of two forms, let alone distinguish between them. The severe form of the trait leads to the individuals being anaemic, and thus likely to die in infancy or adolescence. Many relatives of such 'anaemics' are healthy. However, among these healthy individuals, some show signs of the trait; when the oxygen level is low, for example, high up on mountains, their red blood cells sickle. The distinction between normal individuals, anaemics and 'trait-carriers' (otherwise healthy individuals whose cells can be made to sickle when oxygen is low) was first properly recognized in 1949 by Neel in the United States and by Best in Africa. Neel studied the disease in over 70 families and obtained pedigrees like that shown in Figure 5.

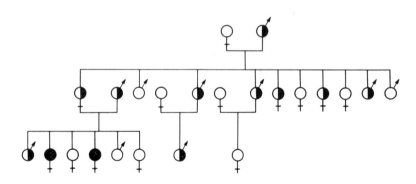

○ normal

◐ non-anaemic trait carrier

● anaemic

Figure 5 A typical pedigree showing sickle-cell anaemia.

ITQ 1 Using Figure 5, answer the questions below (which we first asked in Section 1.1).

(a) Do all members of a particular family inherit sickle-cell anaemia?

(b) Is it only members of one sex that inherit the disease?

(c) Does the trait appear in each generation of the family?

(d) Is there any connection between the disease and the order of birth of the children?

In addition:

(e) Can you see any trends indicating a relationship between those carrying the 'trait' and those with anaemia?

The answers to the ITQs are on pp. 49–51.

So, from the construction of pedigrees of this kind (Figs 3 and 5), reasonable information about the inheritance of the traits can be obtained and they can be shown truly to 'run in families'. The usefulness of such studies in approaching an understanding of inherited traits has been appreciated for some time. In fact, Maupertuis reported in 1768 observations he had made on the family of Jacob Ruhe, a German surgeon, who had six fingers on each hand. Maupertuis examined this trait in Ruhe's parents, his brothers and sisters, and his children. He concluded: 'From this genealogy, which I have followed with exactitude, six-digitism is seen to be transmitted by both the father and the mother'. In sickle-cell anaemia, there is the advantage we have already mentioned of being able to check the pedigrees of several families—not just one, as in the examples of Jacob Ruhe's six fingers or the Habsburg lip. However, well-documented pedigrees of royal families have on occasion provided useful long-term data, another classic example being the inheritance of the disease *haemophilia* among the descendants of Queen Victoria (see Fig. 6 overleaf).

> ITQ 2 Can you notice anything specific about the inheritance of haemophilia that appears not to be true of albinism and sickle-cell anaemia (Figs 3 and 5)?

To make deeper interpretations from pedigrees it is necessary to know the 'rules' for interpretation. These cannot be easily deduced from lineage studies of humans, largely because of the limitations of such studies mentioned in points 1–5 on pp. 8 and 9. The rules for interpretation have come from controlled breeding studies carried out on experimental organisms. We shall first attempt to introduce some of the rules for interpretation and then apply them to the pedigree for sickle-cell anaemia.

1.2.2 Progeny testing

The technique of *progeny testing* involves mating organisms with identifiable differences and observing the inheritance of these differences by examining or testing the *progeny* (offspring) of the matings. Techniques for experimentally breeding animals and plants have been known for thousands of years; they were used to improve or maintain stocks of animals or cereal crops for food. Practical expertise such as keeping stocks 'pure' by preventing matings with other stocks or varieties was common long ago, as witnessed by the Biblical advice: 'Thou shalt not let thy cattle gender with a diverse kind; thou shalt not sow thy field with mingled seed'. However, the deliberate use of experimental breeding as a step towards understanding the mechanisms of inheritance is much more recent. By the eighteenth century, such studies were occasionally undertaken. Réaumur considered making the reciprocal matings of a five-clawed cock with a four-clawed hen and a five-clawed hen with a four-clawed cock. Similarly, he considered reciprocally mating chickens with, and without, a parson's nose.

progeny testing

> QUESTION Can you see any point in doing the matings reciprocally like this?
>
> ANSWER The point, as Réaumur himself envisaged, was that, if the progeny were born without, say, a parson's nose when the parent of one sex had none, but not when the parent of the other sex had none, this would indicate whether the germ originally belonged to the female or the male. This consideration of whether the male or female carried the 'germ' for inherited characteristics was a hotly disputed area in the eighteenth century. Though it is not, in this all-or-none fashion, a controversial topic today (yet what about the German farmer?), the idea of *reciprocal crosses* or matings to reveal relative contributions of male or female to particular traits, is still a useful tool, as you will see in Units 3 and 4. It should be noted, perhaps sadly, that Réaumur never actually mentions doing the experiments he planned nor, of course, any results.

reciprocal mating

11

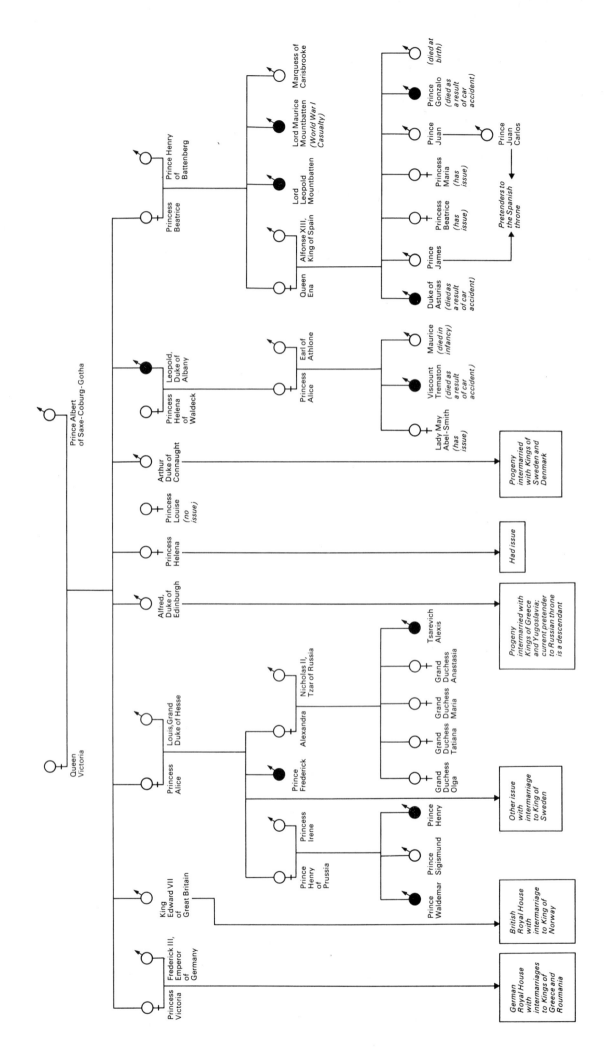

Figure 6 The occurrence of haemophilia among the descendants of Queen Victoria.

1.2.3 Mendelian genetics

By the late eighteenth century and the early nineteenth century, several people had examined inheritance of traits by experimental breedings—mainly in plants. Plants provided useful experimental material: they bred much more rapidly than, say, humans; they were fairly cheap to keep; within any particular species it was relatively easy to obtain varieties with differences in traits such as flower or seed colour; there were no ethical objections to doing experiments on them. It was by experimenting on one such species of plant, the edible pea (*Pisum sativum*), that an Augustinian monk, Gregor Mendel, in the Moravian town of Brno, made his now famous discovery of the basic (Mendelian) rules, or laws, of inheritance although, as you will see, these discoveries made a scientific impact only in the early 1900s, nearly 40 years after they were made (*HIST*).

Mendel's classic paper, *Versuche über Pflanzen Hybriden* (Experiments on plant hybrids), delivered to the Brno Society of Natural Science in 1865, is included as Appendix 1 on pp. 35–47. It is a superb example of how to devise experiments and write an accurate account of methods, results and conclusions. You should read it in association with this Unit and before starting work on *HIST*.

Mendel's experiments are probably, to some degree or other, known to all of you and we shall not attempt to give here a full account of them. The results of his work will permeate this Course as a whole. But in this Unit we would like you to work with some of Mendel's data in the form of an exercise. (You will also have an opportunity to study some of Mendel's conclusions in your home experiment on tomato seedlings.)

First, let us consider some important features of Mendel's experimental system:

1   He worked on the edible pea, *Pisum sativum*, which has the advantage that it is *self-pollinating*. That is, normally, because of the enclosed structure of the flower, a female ovum can be fertilized only by the male pollen from the same flower (Fig. 7). This self-pollination can be called a *self*.

self

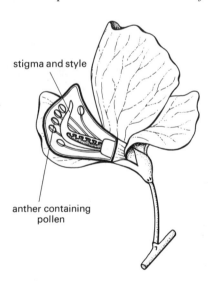

stigma and style

anther containing pollen

Figure 7   The anatomy of the flower of *Pisum sativum*. Some of the keel is cut away, to reveal the male (anthers) and female (stigma) reproductive organs.

The self-pollinating nature of the plant drastically reduces the likelihood of accidental fertilization by pollen from another plant while allowing the experimenter to carry out such a pollination artificially if desired (TV programme 1). A pollination in which ova are fertilized by pollen from other flowers, on the same or a different plant, is called a *cross-pollination* or *cross*.

2   Many varieties of pea were available to Mendel, varying in such traits as seed shape and colour, stem colour, etc. He chose seven traits to examine in detail, some concerned with features of the seed, some with the adult plant, and to begin with examined the inheritance of one readily manageable trait (colour *or* size *or* shape) at a time —an important methodological point. In principle, any characteristic of the plant, at any stage in its development, is suitable for study. Seed characteristics are particularly easy to study as they are always the first visible stages in the development of the progeny plants. The seeds mature soon after the fertilization of the ova and so can be examined sooner than characteristics only manifest in the adult plants, which subsequently develop from these seeds.

3   Mendel made sure that his initial starting varieties (*parentals*) '*bred true*' for the trait under study; that is, when such a plant is selfed, all the resulting progeny are like the parental plant with respect to the particular trait.

Now to the exercise:

Given below is a series of Data and Interpretation Blocks. Read Data Block I. Then choose the interpretation(s) from Interpretation Block I that best fit(s) the data *given up to that point*. Then go on to read Data Block II. Consider the data in Blocks I and II and choose the best interpretation(s) up to that point from Interpretation Block II, and so on. At the end, you should read the Comment Blocks. Should you get stuck on any Data or Interpretation Block, you can, of course, read the Comment Block before you reach the end.

Read Data Block I.

DATA BLOCK I

1   Mendel used two varieties of plant: in one variety the seeds were round, in the other wrinkled (angular). Both varieties 'bred true', that is, when self-pollinated, plants of the round-seed variety gave only round seeds; plants of the wrinkled-seed variety gave only wrinkled seeds.

2   The ova of flowers of a plant producing ordinary round seeds were cross-fertilized with pollen from flowers of a plant producing wrinkled seeds. This cross may be represented thus:

plants grown from round seeds × plants grown from wrinkled seeds

× indicates a cross, or mating.

All the 'hybrid' seeds thus produced (the so-called *first filial generation* or F_1 *hybrids*) were round. Thus:

first and second filial generations

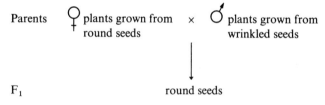

INTERPRETATION BLOCK I

Choose the information that best fits the data in Data Block I.

1   The genetic information specifying round seeds destroys completely the information specifying wrinkled seeds.

2   The genetic information specifying wrinkled seeds is not destroyed; its manifestation is, however, masked by the presence of the genetic information specifying the round trait.

3   The female (the round-seeded variety) carries the 'germ' for inheritance of seed shape.

4   The appearance of the progeny is a result of a mixing of the genetic information from each parent.

Now read Data Block II.

DATA BLOCK II

Mendel also tried the reciprocal cross:

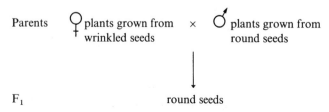

14

INTERPRETATION BLOCK II

Choose the information that best fits the data in Data Blocks I and II.

As for Interpretation Block I, above.

Now read Data Block III.

DATA BLOCK III

1   Mendel took many seeds (all round) of the first filial generation (F_1) of the cross in Data Block I. He grew the plants from these seeds and allowed the plants to self-pollinate. From these he obtained a large crop of *second filial generation* (F_2) seeds. Some of the F_2 seeds were wrinkled (even in the same plant derived from single F_1 seed). That is:

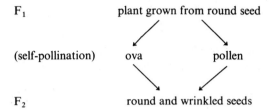

2   Likewise, he produced a crop of F_2 seeds from the plants grown from the F_1 seeds of the cross in Data Block II ($\female$ wrinkled $\times$ $\male$ round). Again, some of the F_2 seeds were round, some were wrinkled. That is:

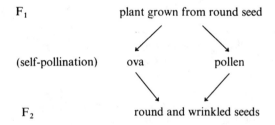

INTERPRETATION BLOCK III

Choose the information that best fits the data from Data Blocks I–III.

1   The genetic information specifying round seeds destroys that specifying wrinkled seeds—hence all the F_1 seeds are round.

2   Seeds resulting from crosses between plants of the round variety and plants of the wrinkled variety contain both types of genetic information, but only the round-variety information is manifest; it is masking the information for wrinkledness. In the F_2, some seeds contain only information for the wrinkled character and hence are wrinkled.

Now read Comment Blocks I–III.

COMMENT BLOCK I

It is not possible at this stage to choose among interpretations 1, 2 and 3—all seem plausible. Interpretation 4 is unlikely as the progeny resemble one parent (the round variety) which is not indicative of 'mixing', though this term is a bit vague anyway.

COMMENT BLOCK II

Interpretation 3 (which incidentally bears on Réaumur's proposed hen experiments) now seems to be ruled out as, if true, all the progeny of the $\female$ wrinkled seeds $\times \male$ round seeds cross should be wrinkled. They are, in fact, all round; 1 and 2 now seem to be the two plausible interpretations.

Interpretation 1 cannot now be held to be true as, if the information for wrinkledness was actually 'destroyed', how was it 'recreated' in such a way that some of the F_2 are wrinkled? Interpretation 2 is possible and, essentially, it is like the conclusion Mendel reached.

Dominance

Mendel considered that factors exist that specify inherited traits, such as seed shape. Such factors can exist in more than one form. So, for example, the factor specifying the shape of seeds can exist in a form specifying round seeds or in a form specifying wrinkled seeds. (In modern terminology we could call such factors *genetic information*.) The hybrid seeds (F_1) arising from a cross between the two plant varieties are like only one parent (round) yet patently contain both forms of genetic information (from the F_2 results). So Mendel considered that one form of a particular trait (here, seed shape) was *dominating* or *dominant to* the other, that is, roundness was *dominant* to wrinkledness and wrinkledness was *recessive* to roundness. When the two types of information were contained in the one organism, only that corresponding to the dominant form was manifest.

dominant
recessive

Phenotype and genotype

As well as the idea of dominance, there is one other important conclusion we can draw from the data presented so far. All the F_1 seeds appear alike (round) and the same as the seeds from which the round-seeded parents grew. Yet, if the parent is self-fertilized only round seeds are obtained; it breeds true. Some progeny (F_2) of a self-fertilized F_1 plant resemble the parent F_1 (they are round), some do not (they are wrinkled) (Data Block III). This means that two organisms (seeds in this instance) can have the same appearance (i.e. they are round seeds) yet apparently contain different genetic information (i.e. some apparently contain genetic information for roundness only, others for roundness *and* wrinkledness). This allows a distinction between the form of a characteristic or trait, which we term the *phenotype*, and the genetic information leading to that trait, the *genotype*. Phenotype and genotype are sometimes used to refer to all the characteristics of the organism and sometimes to refer to the particular characteristics being studied, whether at the level of the properties of cells or of the whole organism. It is usually obvious which usage is being employed but sometimes there may be confusion, of which you should beware. Thus F_1 seeds are of the same phenotype as the dominant (round) parent but of different genotype.

phenotype
genotype

Mendel's simple experiments allow two important concepts to emerge:

1   Genotype as the genetic make-up involved in a particular characteristic or trait, the phenotype*. The same phenotype can result from more than one specific genotype. As Mendel himself realized, 'how risky it can sometimes be to draw conclusions about the internal kinship of hybrids from their external similarity'.

2   The dominance of certain genotypes over others, called recessives.

Mendel's ability to uncover these two concepts undoubtedly depended, among other things, on his choosing a suitable organism, the edible pea, which was experimentally amenable. However, he was not the first person to use such a system. Indeed, Knight in 1799 and Goss in 1824 had both carried out experiments like Mendel's on the edible pea and had, in fact, used similar traits to study. Perhaps we should put down Mendel's discoveries to genius, but if so, it was a genius dependent on another experimental technique that he used that we have hitherto ignored—he counted his plants! This simple expedient, apparently not used by Knight or Goss, enabled Mendel to gain further insight into the nature of heredity. Let us now reconsider the round × wrinkled crosses with some of Mendel's figures attached.

* Note: Mendel himself did not introduce the terms genotype and phenotype, which came into use only about 40 years later, after 1900.

Ratios in F_2

Mendel took 253 F_1 (hence, all round) seeds from the round × wrinkled cross and planted them. On self-pollination of the resultant plants he obtained 7 324 F_2 seeds. Of these 5 474 were round and 1 850 wrinkled. Hence:

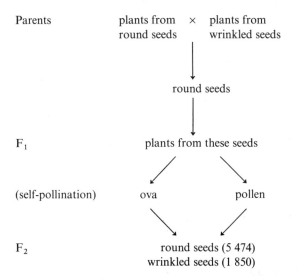

Similarly, he examined a cross between yellow and green seed-bearing plants. All the F_1 seeds were yellow and hence yellow is the dominant form. In the F_2 seeds, 6 022 were yellow and 2 001 green.

Mendel also examined traits exhibited by the adult plants using a cross between a plant containing pigment throughout and a non-pigmented one. This yielded only pigmented plants in the F_1 progeny. In the F_2, 705 plants were pigmented and 224 were not.

> **ITQ 3** Can you see any feature in common with respect to the F_2 progeny in the three crosses we have just described? (Consider the numbers.)

Mendel, in fact, found a ratio close to 3:1 for seven separate pairs of characteristics.

As you will see from Unit 3, and from reading Mendel's paper, he also examined crosses between plants that differed in two or more traits, for example, seed colour *and* seed shape. Thus he cross-bred a plant which was breeding true for yellow round seeds with a plant breeding true for green wrinkled seeds. The F_1 hybrid seeds were grown and the plants they yielded allowed to self-pollinate. The F_2 seeds were examined and counted. Thus:

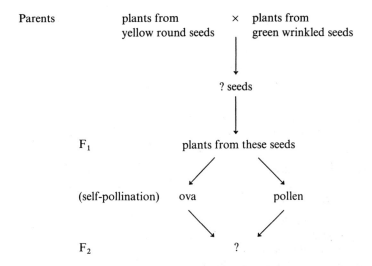

QUESTION What might you expect the F_1 hybrid seeds to look like?

ANSWER They should exhibit the dominant phenotypes—yellow and round.

All the seeds of the F_1 were, in fact, yellow round seeds. The F_2 was composed of four types in the numbers and the proportions shown below:

Phenotypes	Numbers	Fraction of F_2 (approx)	% Total
yellow round	315	$\frac{9}{16}$	56
yellow wrinkled	101	$\frac{3}{16}$	19
green round	108	$\frac{3}{16}$	19
green wrinkled	32	$\frac{1}{16}$	6

Therefore, the ratio of the four types: yellow round (dominant dominant); yellow wrinkled (dominant recessive); green round (recessive dominant); green wrinkled (recessive recessive) $= 9:3:3:1$.

QUESTION For the moment, ignore the numbers. What can you say about the types making up the F_2?

ANSWER There are two 'new' types. That is, two types resemble the parent (yellow round and green wrinkled) and are hence termed *parental types*, but the 'new' types are yellow wrinkled and green round and are *non-parental types*.

The 'new' types in the F_2 appear to arise from a reassortment of the genetic information in the parental types that specify shape and colour.

QUESTION Look at the numbers of the four F_2 types. Now consider each trait separately. What is the ratio of yellow: green and what is the ratio of round: wrinkled?

ANSWER In each case (yellow: green *or* round: wrinkled), it is $\frac{12}{16}:\frac{4}{16} = 3:1$.

So when plants differing in one pair of traits, say seed colour *or* seed shape, are bred the F_2 shows a 3:1 ratio in favour of the dominant phenotype (yellow *or* round). When plants differing in two pairs of traits, say seed colour *and* seed shape, are bred the F_2 shows a 3:1 ratio in favour of each dominant phenotype (yellow *or* round) when considered separately.

QUESTION What does this imply in the terms of the mechanism of inheritance? Choose the best explanation from those given below:

(i) a characteristic is inherited 'in combination' with others;

(ii) the presence of one characteristic (say colour) modifies the inheritance of the other (say shape);

(iii) a characteristic (say colour) is inherited independently of the influence of others (say shape).

ANSWER (iii) seems the best answer. The 3:1 ratio for, say colour, is found in the F_2 irrespective of the presence or absence of differences in shape.

These observations form the basis of Mendel's two laws:

1   Dominant and recessive traits are transmitted independently from one generation to the next, so that neither trait has any influence on the inheritance of the other (principle of independent segregation, sometimes known as Mendel's first law).

2   Inheritance of one gene is not influenced by the inheritance of another gene, and the genes assort independently in successive generations (principle of independent assortment of genes, sometimes called Mendel's second law).

You can now see how the 9:3:3:1 ratio arises. For example, three-quarters of the total F_2 progeny are round and one-quarter wrinkled (3:1 ratio). Likewise, three-quarters are yellow and one-quarter green. So, three-quarters of the round seeds are *also* yellow; that is, three-quarters of three-quarters ($\frac{9}{16}$) of the total F_2 progeny are round yellow seeds. Similarly one-quarter of the round seeds are green, that is, one-quarter of three-quarters ($\frac{3}{16}$) of the total F_2 progeny are round green seeds, and so on. Hence, the $\frac{9}{16}:\frac{3}{16}:\frac{3}{16}:\frac{1}{16}$ (9:3:3:1) ratio. Say, one finds when examining the inheritance of two pairs of traits that four types of progeny occur in the F_2. If they occur in the ratio of $9:3:3:1$ then this is evidence that the inheritance of those traits depends on two pairs of genes, which *assort independently*.

This, as we shall see, is a useful, if not wholly true, concept. It allows one to consider the 'behaviour' of individual *units* of genetic information. That is, one can consider that a 'unit of information' exists for seed colour, a different 'unit' for seed shape, and so on. Furthermore, each unit can be in one of at least two possible forms—yellow or green, round or wrinkled. Nowadays, we refer to a 'unit of information' in the sense implied by Mendel's experiments as a *gene*. So a complex organism has many thousands of 'units' (genes) providing information for various characteristics of that organism. The two, or more, alternative forms of a gene are called *alleles*. (Note: Unfortunately, the terms *gene* or *allele* are sometimes used interchangeably.) Thus the gene for seed colour can be in the yellow allelic form or the green allelic form. This does not mean that the alleles are actually coloured! It is shorthand for either the form leading to yellow phenotype or that leading to the green phenotype. The actual phenotype is, of course, complicated by dominance.

gene

allele

For example, a *diploid* organism, such as the edible pea, contains in its tissue cells two complete sets of genes, one set derived from each *gamete* (egg or pollen in this case) of each parent at fertilization. (A gamete, is *haploid*; it contains just one set of genes.) So, for each trait, a cell in a diploid organism has two copies of the genes (a pair of so-called *homologous* genes), one from each parent. If both copies of the gene are in the same allelic form, the organism is said to be *homozygous* for that gene. If one copy is in one allelic form and the other copy is in another allelic form, the organism is said to be *heterozygous* for that gene.

diploid
gamete
haploid

homologous
homozygous

heterozygous

The phenotype of a homozygous organism (or *homozygote*, as such an organism is called) will, of course, depend on which allelic form is present—a pea homozygous for the green allele of the gene carrying information for seed colour will be green in colour. The phenotype of a heterozygote will be that represented by the dominant allele—a pea heterozygous for green and yellow alleles of the gene carrying information for seed colour will be yellow in colour, yellow being dominant.

> QUESTION What will be the shape of the seed in a plant heterozygous for the genes specifying seed shape?
>
> ANSWER Round. A heterozygous plant will carry one of each allelic form of the gene and 'round' is dominant.

The argument outlined above is really very simple; it is, however, complex to express in words. Fortunately, Mendel was wise enough to adopt a symbolic notation which allows rapid and easy representation. Let us follow the argument through again for seed colour, using the standard genetic notation.

1   All genotypes are represented by italic script.

2   The dominant form of a gene (dominant allele) is represented by a capital letter, the recessive by the corresponding small letter. So, the yellow allele is *Y*, the green allele *y*.

3   The genotype of the homozygous green seed is *yy*, the genotype of the homozygous yellow seed is *YY*, the genotype of the heterozygous seed is *Yy* (or *yY*, the order is irrelevant in this notation).

> **ITQ 4** What would be the phenotypes corresponding to the three genotypes, *yy*, *YY* and *Yy*?

Using the notation, let us now reconsider a cross between two true-breeding plants; one giving yellow seeds, the other green seeds. We can now express both the phenotypes *and* the genotypes.

As the parents are true-breeding for seed colour, they are homozygous.

> **ITQ 5** First, we must satisfy ourselves that this is probably true. To do this, assume that a gamete can be one of two types, that is, any one gamete contains one *or* other of the alleles. Now work through three generations in which the true-breeding plants are crossed to see whether a different phenotype is possible.

19

Now consider the true-breeding yellow × true-breeding green cross.

ITQ 6 Given that all the yellow F_1 seeds are all of the genotype *Yy*, use this notation to derive the genotypes, and hence phenotypes, of the F_2. What are the proportions in the F_2 of the different phenotypes?

1.2.4 Interpreting pedigrees

So now you can see how Mendel's 3:1 ratio in the F_2 fits in with the ideas of dominance, homologous genes, diploid tissue cells, haploid gametes and the fusion of the information in the gametes at fertilization. We have taken the easy route of assuming many of the answers in showing how Mendel's findings fit. Mendel, as you will see when you read his paper, made a brilliant intuitive leap in working from his data. The value of *Mendelian genetics*, as this area is often called, is that it forms a basis for interpreting data from lineages, experimental or natural.

pedigree analysis

ITQ 7 Look at Figure 8 showing a pedigree for albinism, and answer the following questions:

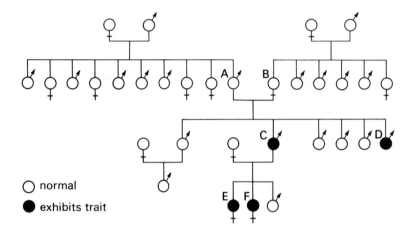

○ normal

● exhibits trait

Figure 8

(a) Do the data confirm that albinism is due to a single gene?

(b) Can you deduce which individual, or individuals, first carry any alleles concerned with albinism?

(c) Is the albino trait dominant or recessive?

(d) Why are the sisters (E and F) albino? Consider the genotype of their mother.

From ITQ 7, you can see that Mendelian genetics does indeed allow interpretation of pedigrees. We cannot 'prove' that the albino trait is definitely inherited in this way, but only that the inheritance implies that the trait is caused by a *recessive allele of a single gene*. In Units 2 and 3, you will see how further application of Mendelian genetics applied to progeny testing tells us even more about the general mechanisms of inheritance and the nature of genes. It is heartening to know that what is true of peas seems to be generally true throughout the plant and animal kingdoms. It is, however, ironic that Mendel's work was more of a 'damp squib' than the 'explosive force' it deserved to be within its own time (*HIST*).

Let us now look back at the inheritance of sickle-cell anaemia. The pedigree chart is shown again below (Fig. 9).

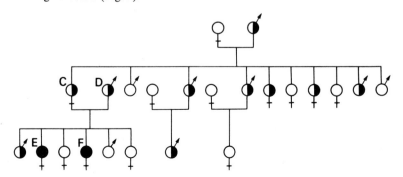

Figure 9 A pedigree showing sickle-cell anaemia.

○ normal

◑ non-anaemic trait carrier

● anaemic

20

The inheritance of sickle-cell anaemia is basically just like albinism. The anaemia occurs in individuals homozygous for a recessive allele, say Hb^S (from Hb for haemoglobin, S for sickle). $Hb^A Hb^A$ or $Hb^A Hb^S$ individuals are not anaemic. (In Hb^A, A stands for the normal wild-type allele.) However, and here is a difference from albinism, $Hb^A Hb^S$ individuals can be seen to differ phenotypically from $Hb^A Hb^A$ individuals. The red blood cells of $Hb^A Hb^S$ individuals show sickling when the oxygen level is low; Hb^A is said to be *partially dominant*. That is, under some conditions, the presence of an Hb^S allele can still be detected in the presence of an Hb^A one. Look at Figure 9 again. Confirm that trait carriers could well be $Hb^A Hb^S$ and anaemics $Hb^S Hb^S$. For example, only where two trait carriers (C and D) mate, does one get anaemic offspring.

partial dominance

1.2.5 The need for statistics

Examine Figure 8 again. The offspring of A and B are pigmented:albino in the ratio $4:2 = 2:1$. The ratio expected is $3:1$. But what would two further children have been, if born? We cannot say. In fact, how many children would one need to examine from a single pair of parents to make sure the ratio for inheritance of a trait was 'correct'? We all know of people who give up 'trying for a boy' after four or five daughters. Yet girls and boys do, overall, occur in roughly equal numbers. So when doing genetic studies, what is the minimum number of progeny to be tested before the actual figures can be trusted? Mendel's ratios are not quite whole numbers, yet he tested hundreds or thousands of progeny. How many progeny must one test before one is sure that the ratio is of biological significance and not due to mere chance? The answer is not absolute. What one can do is establish a set of guidelines based on 'the laws of probability' to get some idea as to how trustworthy a particular set of data is. The need to apply theoretical calculations of probabilities to genetic data has been understood for some time: Maupertuis, during his studies on six-digitism (Section 1.2.1) found only 2 cases in a town of 100 000 people. Allowing for 3 cases that might have escaped his notice, he calculated that 'the probability that this peculiarity will not continue for 3 consecutive generations is 8 million, million to 1' (100 000 to 5, i.e. 20 000 to 1, cubed). Nowadays, more elaborate *statistical tests* have been developed to cope with some of the problems alluded to above. You have ample opportunity to apply such tests (*STATS**) during the Course.

1.2.6 Summary of Section 1.2

1 The main tools of the classical geneticist are lineage or pedigree studies and progeny testing.

2 The data derived from such studies, though numerical, need validating statistically.

3 The ratio of the types of progeny produced by various crosses often, but not always, gives a clue as to the number of genes involved in the inheritance of the traits under examination.

4 The rules for interpreting such studies depend on concepts derived from Mendel. Though not absolute, these rules are broadly true:

(a) 'Units of information' or genes are responsible for individual characteristics.

(b) Identical phenotypes can arise from differing genotypes.

(c) Some phenotypes are dominant to others.

(d) Genes can exist in alternative forms or alleles.

(e) Each gene is represented as a homologous pair in tissue cells; in gametes, only one member of each pair of genes is present per gamete.

(f) 'New' homologous pairs of genes arise following the fusion of gametes at fertilization.

Now try SAQs 1–3 on p. 48.

* The Open University (1976) S299 STATS, *Statistics for Genetics*, The Open University Press. This text is to be studied in parallel with the Units of the Course. From now on we shall refer to it by its code, *STATS*.

1.3 The tools of molecular genetics: 'Frogs and snails and puppy dog tails'

The basic aims of molecular genetics are to identify:

1 The nature of the gene in chemical terms.

2 The molecular mechanisms responsible for transmission of genes from parents to offspring.

3 The mechanisms whereby the information coded in the gene is expressed, that is, decoded to give some product leading to the phenotype.

As the molecular geneticist wants his or her answers in chemical terms, sooner or later he or she must do some chemistry—use the tools of the biochemist. However, much can be learned, and indeed has been learned, about the nature of the gene by using the techniques of classical genetics pushed to their limits. The careful use of progeny testing can give useful data about such problems as the size of the gene, whether the gene is sub-divisible into smaller units of information, and so on. To do such progeny testing, the choice of organism is vital. What is required is a rapidly breeding organism that produces large numbers of offspring and has readily available variants (mutants). The first organism used which fulfilled these requirements was *Drosophila*, the fruit (or vinegar) fly. The choice was initially made in 1909 when Thomas Hunt Morgan, an eminent American developmental biologist, decided that *Drosophila* would be a good organism with which to study the genetics of development. His choice was a brilliant one, and over the next 30 years the 'fly school' provided great insight into the nature of the gene. But the chemical nature of the gene could not be ultimately tackled until a merger of biochemical and classical (formal) genetic techniques occurred to form *molecular* or *biochemical genetics* (*HIST*).

Nowadays, it is commonplace for geneticists to employ all the equipment and techniques associated with the biochemist. The distinction between molecular geneticists, biochemists and molecular biologists is frequently meaningless. However, it was not always so and George Beadle, a founder of biochemical genetics through his classic studies on *Neurospora* (the bread mould) with E. L. Tatum, recounts the problems encountered in the early 1930s when trying to purchase a chemical balance costing $10 while working in Morgan's laboratory: 'We were certain that Morgan would not approve such an expenditure for he knew that geneticists did not need elaborate equipment of that kind'. Times have changed as any glance at a modern department of genetics will prove. The new situation was again brought about by a change of organism; the move was from flies to micro-organisms—fungi, bacteria and bacterial viruses (also called bacteriophages or just phages). These provided organisms suitable for both genetic and biochemical analysis, and the founding of the 'phage school' in the 1940s by Max Delbrück heralded a whole new era of genetic research.

So let us now turn again to sickle-cell anaemia from the point of view of the molecular geneticist armed with the tools developed from the application of classical genetics and biochemistry to micro-organisms.

First, we shall recapitulate the questions:

1 What is the 'molecular reason' for the difference between the normal and the anaemic individual? In other words, what are the molecules that differ from the normal in the anaemic individual?

2 What is the nature of the difference in the genetic information between the normal and the anaemic individual?

3 What 'agents' are responsible for the differences in the genetic information between the normal and anaemic individual?

Question 1 depends on studies of the phenotype, and 2 and 3 necessitate some knowledge of the gene itself. This is not a clear-cut distinction, but a useful one in approaching the topic.

1.3.1 The physiology of sickle-cell anaemia

An understanding of the biochemistry or physiology of sickle-cell anaemia depends essentially on biochemical techniques.

It was well known in the early years of this century that an important physiological role of red blood cells was the transport of oxygen to the tissues. The major protein in red blood cells, haemoglobin, is the substance responsible for this transport. As we have seen (Section 1.2.4), the inheritance of sickle-cell anaemia is consistent with an alteration in a single gene. It was established in the 1930s and 1940s (*HIST*) through the work of Beadle and Tatum on *Neurospora*, that differences in a single gene tend to lead to alterations in a single enzyme. This is now also known to be true for non-enzymic proteins such as haemoglobin. It is, therefore, reasonable to postulate that the single gene mutation in sickle-cell anaemia leads to an altered haemoglobin which is less capable of carrying oxygen than normal haemoglobin. To establish whether this is so or not, requires the techniques of protein chemistry. In 1949 Pauling, Itano, Singer and Wells isolated haemoglobin from red blood cells of anaemics and showed it to differ from normal haemoglobin in some of its physical properties. This difference could well account for the reduced capacity of the red blood cells to carry oxygen. Thus, it seemed that, for the first time, a disease might be explained in terms of an alteration in a single type of molecule that could be examined in the laboratory. However, to the molecular geneticist, the next obvious question was—what causes the difference in the haemoglobin properties? The answer came from the further study of the structure of haemoglobin. Normal haemoglobin consists of four polypeptide chains of two types—two alpha chains and two beta. In 1955, Ingram showed that abnormal haemoglobin of sickle-cell anaemics also comprises four such chains, but in each beta chain the sixth amino acid from one end of the chain is valine instead of the glutamic acid present in normal haemoglobin.

So, in terms of question 1 in Section 1.3.1, this suggests that the anaemic differs from the normal individual in one amino acid in just one type of protein chain. In terms of question 2, this suggests that the information in the gene must, in some way, determine the sequence of amino acids in the protein. Confirmation and expansion of this idea has come from studies mainly on micro-organisms, notably the bacterium *Escherichia coli* and phages that are parasitic on it.

1.3.2 Bacterial genetics and biochemistry of the gene

The usefulness of bacteria and phages will become apparent in later Units. It is not appropriate to deal with those topics here, but we shall briefly summarize some of the advantages of using bacteria for genetic studies, particularly in relation to questions 2 and 3 in Section 1.3.1.

1 Bacteria are haploid. Thus dominance does not mask the genotype (and so complicate its interpretation).

2 Bacteria reproduce by simple cell division. Each of the two daughter cells has the same genotype as the single parent cell.

3 Cell division is rapid. Billions of cells of similar genotype are easily obtained. This is good for biochemical analysis where large numbers of genetically similar organisms are needed.

4 As large numbers of cells are easily obtained, this facilitates production of many mutant forms. This is useful in studying the nature of 'agents' (question 3 in Section 1.3.1) that lead to changes (mutations) in genes.

But how can one do genetics on bacteria? Points 1–4 above do not allow for genetic analysis via progeny testing—for this, 'sex' is necessary, that is, cross-fertilization to produce 'hybrids' (Section 1.2.3).

Fortunately, in the 1940s, it was found that some bacteria have mechanisms for 'fertilization' or at least for transferring genes between individuals. One stage in such a transfer is shown in the photomicrograph overleaf (Fig. 10).

Such transfers have allowed progeny testing. Because of the huge number of progeny that can be tested (the techniques will be described in later parts of the Course), such studies on rapidly reproducing phages have resulted in a much better understanding of the nature of the gene—its size, its substructure, and the nature of the coded genetic information (Unit 6).

This is a very rapid, incomplete survey. In relation to questions 2 and 3 in Section 1.3.1, the techniques we have mentioned (coupled with earlier studies of dividing cells in higher organisms) have led to an understanding of gene structure. We now

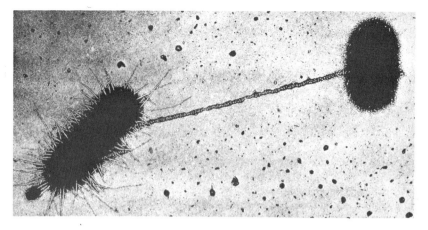

Figure 10 Conjugation between two cells of *Escherichia coli*. During the period when the cells are attached, genes are transferred between them, although they cannot be seen in this photograph ($\times 11\,000$)—see Unit 3.

know that genes are aligned in structures which are visible during cell division in higher organisms—the *chromosomes*. Moreover, we know from studies on bacteria and their viruses that genes are actually composed of deoxyribonucleic acid (DNA), part of the chromosome material. Indeed, we know that the sequence of units (nucleotide bases) comprising the polynucleotide chains of DNA determines the sequence of amino acids in the protein chain. The genetic studies on phage by Benzer and by Crick and their colleagues in the 1950s and 1960s helped towards the first elucidation of which sequences of the nucleotide bases in the DNA corresponded to particular amino acids in proteins—the so-called *genetic code*. However, as you will see in Unit 6, much of the elucidation of the genetic code involved biochemical work on bacteria.

chromosome

genetic code

All this knowledge allows reasonable guesses about the answers to questions 2 and 3, though the gene for haemoglobin has never *itself* directly yielded to such detailed analysis as have those of bacteria or phage. It should, however, be noted that the study of the abnormal haemoglobin of sickle-cell anaemics provides some information about the genetic code and helps confirm that it is fundamentally the same in humans as in bacteria. So we can be fairly certain that the difference between the gene for haemoglobin in the normal individual and that in the anaemic is a change in a single nucleotide base (question 2). As to the agent causing this change (question 3), we can only speculate (Unit 6).

1.4 The tools of population and evolutionary genetics: 'There's a divinity that shapes our ends . . .'

The basic aim of population genetics is to study genetic variation in populations in relation to evolution, by determining the frequency of occurrence of different genotypes in populations of individuals of the same species and by studying the factors affecting these frequencies. To the population geneticist a *population* is all the members of a group of individuals that can potentially interbreed. In practice, the geneticist cannot hope to determine all the different complete genotypes present in any population. This is because of the large number of different genes present in any individual—thousands at least. For example, take a population of diploid organisms in which 1 of 2 possible alleles can occur at each of 100 genes in each organism. When considering the number of theoretically possible different genotypes, remember that any one member of the population can cross-fertilize with any other member. Thus, in this instance, 3^{100} different diploid genotypes are possible.

> QUESTION Why 3^{100}? Why not 2^{100}, as cross-fertilization involves 2 not 3 individuals?
>
> ANSWER The answer depends on the phrase 'diploid genotypes'. At each gene, if one of two alleles (*A* or *a*) is possible, there are three possible diploid genotypes; say, *AA*, *aa*, and *Aa*. For 100 genes, therefore, 3^{100}.

In practice the population geneticist looks at the frequency of occurrence of just a small portion of the complete genotype; the frequency of occurrence of different possible alleles of one or a few genes. How then does he or she go about actually measuring these frequencies? First, the organism, the population and the genes to be studied must be chosen. The population should be one where: (a) it is possible to

24

sample large numbers of organisms so that the results can be tested for statistical significance; (b) the gene studied is one whose phenotypes are easily distinguishable; (c) the gene chosen is one that has detectably different alleles. Point (c) can present problems.

> **ITQ 8** How would you normally expect to detect the presence or absence of a particular allele? Can you see one way in which detection could be complicated in diploid organisms?

Sometimes such problems can be overcome.

1.4.1 Measurement of gene frequency

In the case of sickle-cell anaemia, the frequency of the Hb^S allele can be relatively easily measured. It is fairly easy to examine large numbers of individuals as the phenotype is readily detectable by simple examination of the blood. This can, in fact, be done with simple apparatus, working 'in the field'. Furthermore, because the phenotype of the heterozygote differs from those of the homozygotes (those having the non-anaemic trait as distinct from normals and anaemics), a direct examination of the phenotype gives a direct indication of the presence or absence of the Hb^S allele:

normal phenotype: no Hb^S allele

non-anaemics showing sickling under low oxygen: one Hb^S allele

severe anaemics: two Hb^S alleles

Often, however, the geneticist does not have the advantage of being able to distinguish so readily between individuals having two, one or no copies of a particular allele (see ITQ 8). In these cases he or she needs special techniques to estimate the gene frequencies from the observed phenotype frequencies. Some such techniques will be treated in Units 9 and 10.

1.4.2 Factors influencing gene frequency

There are, as you will see from Unit 9, many factors that can affect gene frequencies. For sickle-cell anaemia the main question is: Why does the Hb^S allele occur at all, or at least, with such a high frequency? (In some areas, up to 40 per cent of the population carry an Hb^S allele.) Because many individuals homozygous for Hb^S die in infancy and therefore do not reproduce, we would expect the frequency of the Hb^S allele in the population to fall fairly rapidly from generation to generation. Yet it seems to be maintained at a fairly steady and surprisingly high level in some areas of the world. As we asked earlier (Section 1.1.3): *Why is it so maintained, why in those particular geographical areas and how are the populations in different areas connected?*

A clue to the answers came in the 1950s when Allison found a high incidence of the Hb^S allele in humans in geographical areas with a high incidence of malaria. Further studies showed that the possession of the phenotype arising from the Hb^S allele actually confers some resistance to the disease. Thus, in this respect, anaemics ($Hb^S Hb^S$) and non-anaemic trait-carriers ($Hb^A Hb^S$) would appear to have a higher chance of survival to maturity than normal individuals ($Hb^A Hb^A$), and hence a high chance of passing on the Hb^S allele to the next generation. So where malaria is common, the possession of the Hb^S allele confers an *advantage*. But the Hb^S homozygotes are very anaemic and tend to die young, so the Hb^S heterozygotes, the non-anaemic trait-carriers, are the most favoured group. This is a case of *heterozygous advantage*, the heterozygote being relatively better fitted for survival than either of the two homozygotes—*but only where malaria is common*. This is an example of an interaction* between the genotype and its environment (see Section 1.7).

heterozygous advantage

So the African Negro population tends to have a high frequency of the allele, as it lives in areas with a traditionally high incidence of malaria. But what about such a population transported to a non-malarial area? Does the allele survive at the same

---

* In most of this Course, we use the word interaction in a general sense, without defining it precisely. However, there is a precise and limited way in which the word is used in biometrical genetics (Units 11–13), where a factor known as genotype–environment interaction is introduced into equations. When we mean the term in the specific sense, we shall make it clear.

level in subsequent generations? We can ask this question of the American black population, largely descended from African slaves.

> **ITQ 9** (*Objective 6*) If the ideas outlined above, stemming from Allison's discovery, are correct, what would you expect about the frequency of the Hb^S allele in the modern American black as compared with the modern African Negro?

Studies over the last few years have confirmed a lower incidence of the Hb^S allele in the American black than in the African Negro. Some of the lower incidence almost certainly results from interbreeding between the original Negro population and the American whites. But as well as this, it can be shown that selection against the Hb^S allele has been a significant factor.

So we can conclude that one factor affecting the frequency of the Hb^S allele in the various populations in different geographical areas is the occurrence of malaria in those areas. In the case of the Hb^S allele, the main factors involved in maintaining a balance of the frequencies of the allele are a *selection pressure* for individuals carrying the allele where malaria occurs versus the deleterious effects of anaemia under conditions of low oxygen.

selection pressure

Very few other alleles lend themselves to such clear study because, unlike Hb^S in the homozygous state, they have no obvious lethal effect, and therefore it is less easy to identify the factors that affect the frequency of the allele in a population. These factors and the methods used to study them will be considered in Units 9 and 10. Just for now, we name some factors that can affect gene frequencies:

1 Selection pressures for and against individuals carrying a particular allele.

2 Isolation of groups within the population from other groups by, say geographical or social barriers. This can give rise to inbreeding (breeding between closely related individuals).

1.4.3 Evolutionary genetics

We have already spoken of selection pressures when discussing population genetics. As you know, natural selection is the motive force of evolution and, for this and other reasons, the interests of population geneticists and evolutionary geneticists are virtually the same. The main problem for the evolutionary geneticist is that he or she is essentially a 'historian', trying to reconstruct what has happened over millions of years and, in particular, to determine how gene frequencies have changed in populations with time, how these changes have affected evolution, and how they lead to new species, etc. You can see from the last Section how sickle-cell anaemia provides the evolutionary geneticist with a 'model' of how *natural selection* can operate to favour one group over another.

natural selection

To go further, the evolutionary geneticist has to make estimates of, for example, the rate of evolution, based on such things as data from population studies showing how rapidly gene frequencies can alter; the comparison of genes in one species of organisms with those of another; mathematical models to interpret and simulate such data. We shall not deal further with these techniques here, but their use will become apparent in later Units.

1.5   Are there other important approaches to genetics?

As we explained in Section 1.1, the division of genetics into three main areas (formal, molecular and population) is somewhat arbitrary. Nowadays the divisions are very blurred and individual practising geneticists would certainly not want to be 'pigeon-holed'. A molecular geneticist will use 'classical' techniques, a population geneticist needs to know the rules of the inheritance of traits and where possible the molecular basis of inheritance. All of genetics is relevant to any geneticist, which should already be clear to you from the sickle-cell anaemia example—and will be even more so by the end of the Course. However, there is also one other important area of genetics, which we intend to deal with later in the Course, *developmental genetics* (Unit 8), currently a field of great interest in biology.

developmental genetics

Developmental biology is concerned with the processes involved in the change from a fertilized egg to an adult animal or plant. It is reasonable to suppose that a genetic component is involved in these processes—involved, that is, in two ways. First, the genetic endowment of the single fertilized egg cell obviously determines that the future adult organism broadly resembles its parents: a frog's egg gives rise to a frog, a rabbit's egg to a rabbit, and so on. Secondly, the genetic component is also presumably involved in the order of development, that is, the well-regulated sequence of changes that occur from egg to adult. This second involvement we can term a *genetic programme for development*. We are not, of course, implying that the genetic endowment of the fertilized egg is the sole arbiter of development; the environment in which development occurs is also very important. The interaction between environment and genotype is crucial to development.

genetic programme for development

The developmental biologists of the late nineteenth century attempted to unravel the mysteries of development by interfering with the environment. It was appreciated soon after the rediscovery of Mendel's findings that an equally valid approach to developmental biology would be to investigate the nature of the genetic programme for development. To this end, in 1909 Morgan embarked on an investigation of the genetics of *Drosophila* (*HIST*). As we now know he was brilliantly 'side-tracked' into a long-term investigation of the nature of the genetic system itself and never got to grips with the developmental programme, as such. However, thanks to his efforts and the efforts of those who followed, some 60 years later, research on the genetic programme for development (developmental genetics) is in full swing. As you will see from Unit 8, it is one of the most exciting areas in the whole of biology.

1.6 The right organism: the home experiments

You will have already become aware of how important it is to choose the right organism when embarking on a particular line of research. For example, although peas were useful for defining the basic laws of inheritance, a more rapidly breeding organism, the fruit-fly, was needed to learn more about the nature of the gene. Similarly, even more rapidly breeding organisms, such as bacteria, were needed to enable an understanding of the chemical nature of the gene. However, we do not wish to give the impression that scientists are always able to choose the right organism in advance of their studies; frequently, our rationalizations of why an organism is particularly useful are done with all the benefit of years of hindsight. Sometimes the reasons for originally choosing a particular organism are apparently less than carefully planned—as George Streisinger, a molecular biologist, recounts in his explanation of why he chose to work on phage:

> The work with plant viruses and plant tissue cultures was very fruitful at first. Harry Rubin and I shared a lab; Rubin worked on Rous-sarcoma virus and the by-products of his experiments were chickens (minus a wing, or leg perhaps, but still very good to eat), while I produced a vast excess of coconuts whose milk seemed to be essential for growing plant tissue cultures. Harry Rubin ate part of my coconuts, my family ate Rubin's partial chickens; an idyllic relationship which lasted for several months. After that, alas, my family got tired of eating chicken, I got tired of plant viruses not forming plaques and, besides, there seemed to be so many appealing experiments to be done with phage.

We hope we have shown foresight in selecting for your home experiment organisms that will yield some insight into the techniques and limitations of classical and molecular genetics (if not your next dinner!).

1.7 How phenotype depends upon both genotype and the environment

You have seen already something of the range of interests of a geneticist. Basically those interests are in one way or another concerned with describing the characteristics (the phenotypes) of individual organisms or groups of organisms in terms of inherited factors—in terms of their genotypes. As you know, the same phenotype can arise from different genotypes—yellow peas can be homozygous *or* heterozygous for a dominant allele conferring yellowness. Does a complete description of an

interaction between genotype and environment

individual organisms' genotype allow, in principle, a complete prediction of that individual's phenotype? Put another way, given two individuals of identical genotype, are their phenotypes necessarily identical? The answer, to some extent, depends on one's definition of 'identical' and 'phenotype'. For example, to the molecular geneticist, the phenotype corresponding to the $Hb^S Hb^S$ individual is the abnormal haemoglobin. To a doctor, the phenotype of the same individual is that he is anaemic or, in severe cases, has circulatory defects. Each of the phenotypes is a consequence of the same genotype, $Hb^S Hb^S$, and there is a hierarchical connection between them: the abnormal haemoglobin results in reduced oxygen levels in the red blood cells; these sickle and do not flow readily through the capillaries. This can lead to blockages, and hence circulatory problems.

So the phenotype can be considered at several levels. Let us for the moment define phenotype broadly as *any* characteristic of an organism dependent on genotype, and return to the question posed above: 'Given two individuals of identical genotype, are their phenotypes necessarily identical?' The answer is frequently, *no*! Take, for example, two pea plants, each grown from yellow peas of identical genotype, homozygous for yellowness. If one plant is grown in one type of soil and the other grown in another type of soil differing in, say, iron content, it is quite likely that the F_1 generations resulting from self-fertilization of these plants will differ to some extent in the degree of yellowness of the peas. Even on the same plant, not all peas will be truly identical in colour; there will be some variation. Such variation is, of course, due to environmental factors. So the phenotype of an organism or the phenotype of a particular feature of that organism, is a result of an interaction between the genotype of that organism and its environment. Likewise, it can therefore sometimes be possible to arrive at the same phenotype from two different genotypes because of different environments (for example, a pea that is genetically green may be yellow because of a lack of certain minerals in the soil). So, when examining the inheritance of traits, care must be taken to keep relevant environmental factors under careful control. The important point to remember is that the genotype is inherited, not the phenotype (Section 1.2.3). So the next generation from any organism depends on the genotypes of the parents (not on their phenotypes) and its own environmental conditions.

To take some more examples of how phenotype depends on both genotype and environment:

1   Many congenital defects in man (congenital merely meaning 'evident at birth') are due to genotypic effects, for example, sickle-cell anaemia. However, other congenital defects may be due to damage to the fetus while in the uterus and may not be genetically determined; thalidomide-produced deformities are instances. In these cases, just one component, genotype *or* environment, can be defined as the cause of the phenotypically evident defect.

2   Again defining phenotype broadly, an adult human is a result of its genotype and its environment as a child—food and health-care being of obvious importance. For example, what would genetically be black hair in many Africans is rendered red by the nutritional disease kwashiorkor, which occurs through an inadequate supply of protein during childhood. Note that this depends on an *interaction* between genotype and environment. A genetically blond child would not gain red hair as a result of kwashiorkor.

These interactions are of a relatively simple, one-way, nature, the resulting phenotype being definably affected by the genotype and/or environment. But, frequently the interactions are more complex and the contribution of the genetic or environmental components hard to sort out. This is particularly true of humans in respect of 'behavioural phenotypes'. Even a simple example illustrates some of the complexities: red–green blindness (the inability to distinguish red from green) is a genetic trait, a form of which occurs in male humans due to a single recessive gene transmitted via (but very rarely in) the female line, like haemophilia. (How it is transmitted will be apparent from Unit 2.) But how do we normally define the phenotype of such a man? The quick answer is, 'he's colour blind', but what does this mean? His colour blindness is only manifest, or relevant, when defined in a particular behavioural way. That is, in some environments, he behaves as would a man with normal, full-colour vision; in others he may show a different behavioural phenotype. For instance, he can play billiards without any disadvantage at all, but when playing snooker, he needs a benevolent opponent. (For non-addicts, billiards depends, among other skills, on distinguishing red from white balls; snooker means recognition of white, yellow, green, brown, blue, pink, black and red balls.) So when discussing complex behavioural phenotypes, particularly in humans, it is necessary to define, not just their supposed genotype, but the environment in which one is considering that genotype. The problem of arriving at these two definitions is often very difficult, as you will see (Section 1.9 and Unit 15).

1.8 Genetics and society

Knowledge gained from studies in genetics is often of application to human society, the use of blood tests in paternity suits (Section 1.0) being a trivial example. In Units 14 and 15 and in *HIST*, we shall look at some such applications in detail. For now, we shall briefly indicate some of the areas in which the knowledge of inheritance has been applied to humans and their societies and other areas where it might be applied in the future.

1.8.1 Animal and plant breeding

Humans have been selectively breeding plants and animals for argicultural purposes for thousands of years. Though often a slow process, selective breeding has been very successful in producing crops fitted to local environments, cattle for high milk or beef yield, sheep for good quality wool, and so on. Until very recently, all this was achieved with no knowledge of genetics other than what the breeder's eyes told him—'like begets like'—choose the characteristics you want in the parents, and the offspring have a fair chance of inheriting them. So where can our knowledge of the laws of inheritance help? There are probably two general ways:

1 By knowledge of the actual degree of genetic dependence of a particular trait (i.e. whether it is dominant or recessive, whether other genetic traits are closely linked or not, what proportion of individuals should be affected) and by measurement of quantitative aspects (Unit 13), genetics can, perhaps, improve the 'fair chance' referred to above.

2 A detailed knowledge of the molecular nature of the gene and the ability to manipulate cells of higher organisms as we would bacteria, may eventually allow the possibility of '*genetic engineering*', that is, the artificial production of new varieties of animals or plants by altering the genotype of existing varieties without the slow hit-and-miss of selective breeding.

We shall return to these possibilities later in the Course in Units 12, 14 and 15.

1.8.2 Medical applications

A knowledge of which diseases have some genetic basis can often help to prevent, cure or alleviate symptoms of the disease. For example, humans have several different classes of protein on the surface of their red blood cells. These classes of protein determine the various classes of blood group. One classification of blood groups is the rhesus system. At its simplest, people are either rhesus positive or negative—they either have the rhesus factor in their red blood cells or not. The two types of blood are incompatible. Transfusion of rhesus-positive cells into a rhesus-negative person results in the production by the recipient of antibodies (specific proteins that identify and 'attack' foreign proteins) against the rhesus factor; the donor blood cells are consequently destroyed. Care in typing blood in hospitals easily avoids such mishaps. The rhesus system is genetically determined. We can represent the genotype of the rhesus-positive as Rh^+, the rhesus-negative as Rh. Consider what can happen when an $RhRh$ woman mates with an Rh^+Rh^+ or Rh^+Rh man. The resulting embryo can be Rh^+Rh (the actual genetics of the system is complex and will not be discussed). If some fetal blood mixes with the mother's blood, she will become immunized and produce anti-rhesus-factor antibodies. Usually these are produced in small amounts for a first Rh^+ pregnancy and no harm results. However, in subsequent pregnancies an Rh^+ fetus gives rise to large amounts of anti-rhesus-factor antibodies, and can result in the production of a badly jaundiced child suffering from haemolytic disease, or even death of the fetus (see Fig. 11). This can now be relatively easily avoided by carrying out a blood transfusion on the unborn or new-born child, or delivery by Caesarean section, or prevention of antibody attack before damage occurs. A knowledge of the genetics of the rhesus system is very useful here as it allows the doctors

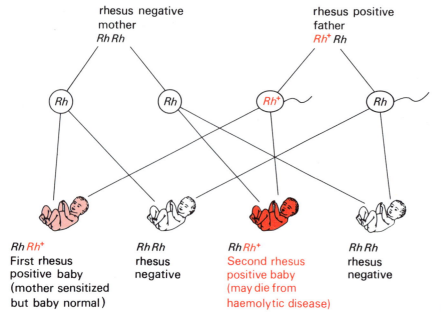

rhesus negative mother
Rh Rh

rhesus positive father
Rh⁺ Rh

Rh Rh Rh^+ Rh

Rh Rh⁺
First rhesus positive baby (mother sensitized but baby normal)

Rh Rh
rhesus negative

Rh Rh⁺
Second rhesus positive baby (may die from haemolytic disease)

Rh Rh
rhesus negative

Figure 11 The rhesus system and pregnancy.

to test for the prospective parents' blood groups and to take measures to ensure the child's safety. With the development of techniques for testing for rhesus and other factors in early-stage fetuses by sampling blood or fluid from the fetus *in utero* (amniocentesis) further predictions of genetic or other anomalies may soon be obtained routinely. There are, however, ethical as well as medical or genetic problems in such developments.

1.8.3 Human breeding

By testing for the blood groups of prospective parents, one can be forewarned of danger to a fetus. Extending this idea, one can, in principle, with the relevent knowledge of the rules of inheritance, predict the likelihood of particular individuals' passing on particular traits that they carry. What one does with this knowledge is a moot point. At one end of the scale there is *genetic counselling*, advice intended to help individuals with some defect that they are afraid they may pass on to their offspring. Often the matter that worries them is some trait that is known not to be heritable and immediate assurance can be given. Where the trait is genetic, and if the knowledge is available, individuals are told what the mathematical chances are that children will inherit the trait. The decision about what to do with this 'new knowledge'—whether to have children or not—is, in principle, left to the individuals seeking the information, but sometimes advice is given. However, such advice would tend to be at the level of individual families. At the other end of the scale might be policies of *enforced genetic 'advice'*, which would mean the prevention of certain individuals from having children and/or the encouragement of others to breed. Such *eugenic* policies have been proposed in the past, and are sometimes put forward today. Whether such policies are ever justified is a moral and social issue rather than a scientific one. To see why this is so, we discuss the history of eugenics, and the genetic arguments about whether eugenic proposals might work, in *HIST* and Unit 15.

genetic counselling

eugenics

You can see that, having started this Section on 'genetics and society' from the standpoint of how genetic knowledge affects society, we are now in areas where social mores and prejudices influence what is done with such knowledge. The social pressures also affect what actual genetic experiments are done and, frequently, how the knowledge is interpreted. Genetics and society interact. Such interactions are two-way, but we shall attempt, as we did (almost successfully) up to Section 1.8.3, to look at just one direction at a time. In the next Section we turn the phrase around—society and genetics.

1.9 Society and genetics

Any scientist is an inheritor or victim of the social values of his time. His research, his discoveries, his ideas are all subject to the social system in which he lives whether he is fully aware of it or not (and note that it *is* generally *he* and *his*, not she and hers, that we are discussing in science as practised today!). Often his discoveries radically affect society itself. The effects of society on his work are usually less obvious and more indirect, although as research costs money, and the money comes almost exclusively from the state or industry, it is obvious that certain types of work are more likely to be supported than others—especially if they can be seen to contribute to military or industrial developments. Nuclear physics was regarded as a typical 'boffins'' ivory tower until Hiroshima; now popular newspapers discuss the rights and wrongs of fusion power, reactors and such like. Molecular biology seemed similarly to be a typically esoteric subject until there was a suggestion (still unproven) that certain viruses might be linked with cancer, and until research into biological warfare got under way. Today everyone is aware of possible biochemical cures for cancer—money pours in for cancer research, from the state and from charities, and hence more biological research gets done. And so on.

The study of heredity, that is genetics, has always been of great general interest and the social values and prejudices of the time have greatly and obviously influenced the direction of research and thought. 'Like begets like' is known to everyone by common observation, and anything apparently deviating from such observation is suspect. The suspicion is dealt with either by accusation of impossibility (as in the case of the German farmer) or by rationalization by natural philosophers within the common folklore. Thus, Malebranche in his 1700 tract, *Recherche de la Verité*, writes:

> . . . a woman gazed too long at the picture of St. Pius, whose Canonization was being celebrated, and she gave birth to a child who perfectly resembled the picture of the Saint . . . This was seen by all Paris, as well as by me, as it was preserved for quite a long time in alcohol!

Even when new techniques appear to offer opportunities to dispel prejudices, prejudices influence the use and interpretation of technical advances. Thus, when Hartsoeker in 1694 examined semen in an effort to settle a debate (much loved in the seventeenth and eighteenth centuries) about whether the male or female carried the 'germ' of inheritance, he observed that each sperm 'contained' a little male or female animal of the same species hidden under a tender and delicate 'skin' and, significantly, adds that fertility being thus attributed to the male 'is more in keeping with his dignity'.

The influences of society on genetic studies are analysed in *HIST*, but as the examples we have given here may seem quaint, let us briefly deal with a few more up-to-date interactions between society and genetics.

Mendel's work

Mendel's work was not widely known, nor was its significance realized till well after his death. Some of the reasons why his work was not appreciated will be dealt with in *HIST*, but among them were the originality of his ideas and the lack of widespread communication between scientists.

Evolution versus the Bible

In the late nineteenth century, great controversies raged between supporters of Darwin's theories of evolution (notably Thomas Huxley) and eminent Churchmen such as Bishop Wilberforce. Even now, strongly religious groups are trying to introduce legislation in California to ensure that teaching in schools gives equal weight and probability to 'creationist' ideas derived from the Bible and evolutionary theory. This may seem a hangover from Victorian religious fundamentalism. However, more up-to-date examples of the effect of social values on genetics, and on geneticists themselves, are available.

Racism and genetics

The ideas embodied in Mendel's laws and the theory of evolution have been frequently 'used' in an attempt to provide support for certain ideological viewpoints and political prescriptions. Thus, in the 1920s in the United States, apparently 'scientific' genetic data were used by eugenicists to support their policies of 'white supremacy', the limitation of immigration from Southern and Eastern Europe and even sterilization programmes (*HIST*). Similarly, in Germany in the 1930s some geneticists argued that their research supported Nazi ideas of 'racial purity'. Conversely, where particular genetic ideas have run counter to the political climate, they have been ignored, or indeed vilified. Notably, in the USSR in the late 1930s and 1940s, 'Mendelism' and its development were officially condemned. The argument against Mendelism was led by a Russian agronomist, T. D. Lysenko, and so effective was he that orthodox Mendelian studies and those who wished to carry them out were outlawed (*HIST*). Once again, this had far-reaching consequences—not just to genetics—and the techniques of the mass media were employed to show the apparent relationship of Mendelism with capitalism and its links with racism (Fig. 12).

Figure 12 Four views of Western genetics from the pen of the famed Soviet caricaturist Boris Efimov, which appeared in the 1949 *Ogonyok*, a popular journal, to illustrate an article entitled 'Fly-lovers—Man-haters'. The inscription on the flag reads 'The banner of pure science'.

Genetics and intelligence

According to a journalist, Robert Conquest, in an article in the *Daily Telegraph* of 20 April, 1974, headed 'Common Sense on Colour Blindness':

> ... Apart from colour, there are the genuine ethnic divisions; the Caucasian, Mongolian and Negro stock, plus the smaller Australian and Bushmen races. While we assert the brotherhood of man, we need not deny its diversity.

> It is absurd to expect, or to think it desirable, that races which have bred so different physically over such a long period are bound to be identical psychologically. A year or two ago, one of the science magazines conducted behaviour tests on four breeds of dog. They acted quite differently in each situation—my own favourite, the hound, coming out particularly in matters of self-control.

> And though these differences arose through artificial breeding, this was over a far shorter period than the natural breeding which differentiates humans. "Liberals" have always been unwilling to face this, and, in particular, the now undeniable evidence which shows that whatever environmental allowances are made, the black perform [sic] differently in certain IQ tests from the white, the Eskimo or the Chicano.

There are a number of underlying genetic assumptions made in this extract, including; (a) that human 'race', as defined by skin colour, facial characteristics, etc., has a distinct genetic meaning and (b) that there is a genetic component to intelligence, at least as measured by IQ tests, which may be related to the 'racial' differences.

We examine the genetic basis—or lack of it—for such assumptions in Units 14 and 15. For the moment, however, we would like to ask how much different this writing is from an earlier one, which appeared in the *Sunday Times Magazine* as a caption to a photograph of some Pigmy tribesmen visiting London in 1905:

Surely extremes met when the little folk from the heart of the Ituri Forest, in Central Africa, mixed with the Members on the Terrace of the House of Commons. They are supposed to be of the lowest type, mentally, as well as the smallest, physically, of the human race. What did they think of the greatest Legislature of the world? What dim conception did they form of its purposes and of its work? Probably they said it is a good place for honey and lime-juice, the two things of civilisation for which they cultivated the keenest zest.

One substantial difference is that where the 1905 caption says 'They are supposed to be · · ·' the recent article says 'Now undeniable evidence · · ·', this, apparently, being scientifically solid genetic data.

An important aim of our Course is to enable you to assess this 'undeniable evidence', and to be able to distinguish scientifically valid statements from mere ideological writing masquerading as science from whatever quarter—whether 'liberal' or 'reactionary'—such writing may appear.

The point is that the issues are much more complex than they are often argued to be. When one deals with human behaviour and asks genetic questions about it, one is (as you will see in more detail in Units 14 and 15) attempting to analyse a behavioural phenotype which depends upon the interactions between genotype and environment. To illustrate the difficulties of investigations of this sort, let us briefly return to our red–green blindness analogy (Section 1.7). The inability to play, unaided, a good game of snooker is no serious handicap—you just buy your own beer more often. However, what about driving cars? Should red–green colour-blind people be allowed to drive? Is it socially responsible to allow them to drive? Consider what happens when you, as a driver with full-colour vision, approach traffic lights—you see whether they are red or green and take appropriate measures. The red–green colour-blind driver relies on the position of the lights, *not* on their colour (the top light means 'stop', the bottom one 'go').

What happens when it is foggy? The normal driver can still see red or green, admittedly at a lesser distance. However, the colour-blind driver cannot distinguish between them *until he can see the outline of the array of lights*—at a considerably lesser distance. In a fog, the driver who cannot distinguish red from green is therefore considerably more dangerous at traffic lights. His genotype renders his phenotype in fog less socially acceptable. What should society do? There are two extreme views:

1   We should test all drivers at the time of their driving test for colour vision, and fail all with red–green colour blindness. (To save cost, one could test only men—a victory for women drivers?)

2   We should re-design traffic lights. For example, make the red (stop) light a flashing one and the green (go) light a static one; or have two lights for 'stop' and a single light for 'go'.

Even in a simple case like this, where one has the advantage of being able to distinguish between the genetic and the environmental component (which includes defining the interactions between them), there are *at least* two completely different ways of 'correcting' the behavioural phenotype (the ability to distinguish 'stop' from 'go'). Where one cannot easily distinguish the component, and indeed where such a distinction may be completely without meaning, caution should be exercised before concluding that altering *either* the genotype *or* the environmental component is the only way to 'correct' the phenotype. It is the interaction that is altered when either of the components is altered; it might, therefore, be possible if one is attempting to derive a policy conclusion from a scientific observation, to reach, as in the case of colour blindness, the same result in any one of many ways.

1.10   The logic of the Course: 'The best laid schemes o' mice an' men . . .'

In this Unit we have adopted a certain structural logic, placing classical genetics and the interactions between genotype and environment as our lynch pins. This was done merely as a way of introducing some of the interesting ideas of genetics and to give a taste of some of the methods involved. There is no single logical way of teaching genetics. In fact, the Course Team spent a lot of time considering various alternatives—should we start historically, start with DNA, consider populations first, and so on? We had obviously to decide on a particular sequence. The one we chose starts in Unit 2 with a consideration of genes and chromosomes. We hope it makes logical sense to you—and that you enjoy the Course!

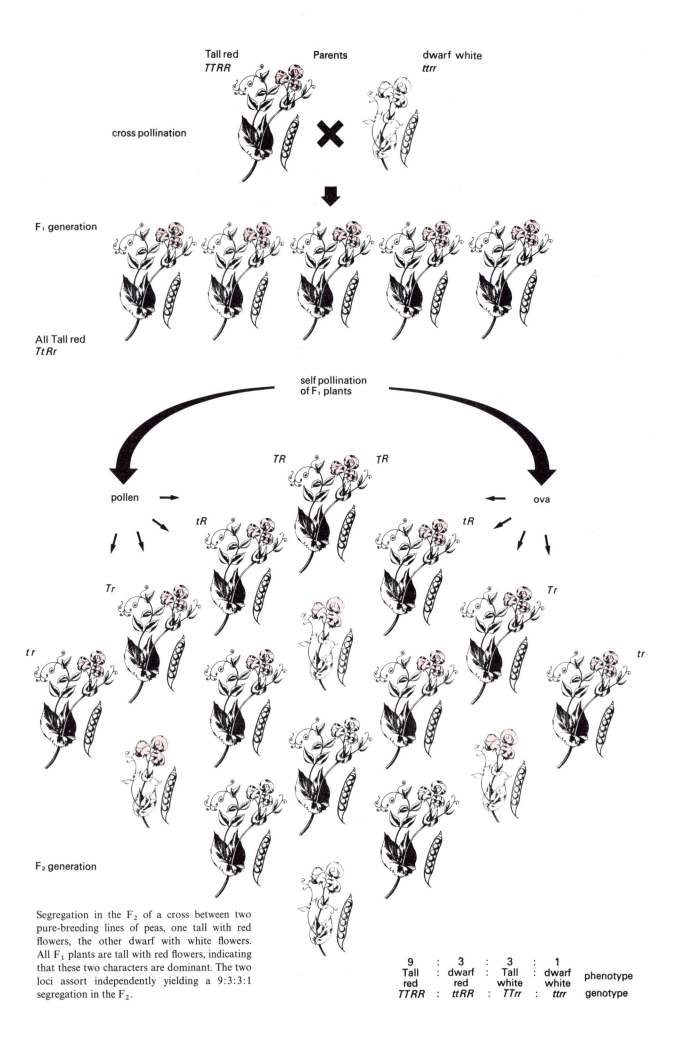

Tall red
TTRR

Parents

dwarf white
ttrr

cross pollination

F₁ generation

All Tall red
TtRr

self pollination
of F₁ plants

pollen

TR   *TR*

tR   *tR*

ova

Tr   *Tr*

tr   *tr*

F₂ generation

Segregation in the F₂ of a cross between two pure-breeding lines of peas, one tall with red flowers, the other dwarf with white flowers. All F₁ plants are tall with red flowers, indicating that these two characters are dominant. The two loci assort independently yielding a 9:3:3:1 segregation in the F₂.

9	:	3	:	3	:	1	
Tall red		dwarf red		Tall white		dwarf white	phenotype
TTRR	:	*ttRR*	:	*TTrr*	:	*ttrr*	genotype

Appendix 1: Mendel's paper, Experiments on plant hybrids

Our translation of Mendel's original paper, Versuche über Pflanzen Hybriden, published in 1865 in the *Abhandlungen des naturforschenden Vereines in Brünn*, Bd, iv, is closely based on the translation of 1901 by the geneticist William Bateson (see *HIST*) in the *Journal of the Royal Horticultural Society*, Vol. xxvi.

For interest, we include verbatim the Introductory Note written by Bateson, which places the discovery of the paper in its historical context. We have also retained his footnotes and use of italics.

Experiments on plant hybrids, by Gregor Mendel

Introductory note

The original paper, of which the following pages are a translation, was published by Gregor Mendel in the year 1865 in the 'Abhandlungen des naturforschenden Vereines in Brünn', Bd. iv. That periodical is little known, and probably there are not half a dozen copies in the libraries of this country. It will consequently be a matter for satisfaction that the Royal Horticultural Society has undertaken to publish a translation of this extraordinarily valuable contribution to biological science.

The conclusion which stands out as the chief result of Mendel's admirable experiments is of course the proof that in respect of certain pairs of differentiating characters the germ-cells of a hybrid, or cross-bred, are pure, being carriers and transmitters of either the one character or the other, not both. That he succeeded in demonstrating this law for the simple cases with which he worked it is scarcely possible to doubt.

In so far as Mendel's law applies, therefore, the conclusion is forced upon us that a living organism is a complex of characters, of which some, at least, are dissociable and are capable of being replaced by others. We thus reach the conception of unit-characters, which may be rearranged in the formation of the reproductive cells. It is hardly too much to say that the experiments which led to this advance in knowledge are worthy to rank with those that laid the foundation of the Atomic laws of Chemistry.

To what extent Mendel's conclusions will be found to apply to other characters, and to other plants and animals, further experiment alone can show. Though little has yet been done, we already know a considerable group of cases in which the law holds, but we also have tolerably clear evidence that many phenomena of cross-breeding point to the coexistence of other laws of a much higher order of complexity. When the paper before us was written Mendel apparently inclined to the view that, with modifications, his law might be found to include all the phenomena of hybridization, but in a brief subsequent paper on hybrids of the genus Hieracium* he clearly recognized the existence of unconformable cases.

Nevertheless, however much it may be found possible to limit or extend the principle discovered by Mendel, there can be no doubt that we have in his work not only a model for future experiments of the same kind, but also a solid foundation from which the problem of Heredity may be attacked in the future.

It may seem surprising that a work of such importance should so long have failed to find recognition and to become current in the world of science. It is true that the journal in which it appeared is scarce, but this circumstance has seldom long delayed general recognition. The cause is unquestionably to be found in the neglect of the experimental study of the problem of Species which supervened on the general acceptance of the Darwinian doctrines. The problem of Species, as Gärtner, Kölreuter, Naudin, Mendel, and the other hybridists of the first half of the nineteenth century

conceived it, attracted thenceforth no workers. The question, it was imagined, had been answered and the debate ended. No one felt any interest in the matter. A host of other lines of work were suddenly opened up, and in 1865 the more vigorous investigators naturally found those new methods of research more attractive than the tedious observations of the hybridizers, whose inquiries were supposed, moreover, to have led to no definite result. But if we are to make progress with the study of Heredity, and to proceed further with the problem 'What is a Species?' as distinct from the other problem 'How do Species survive?' we must go back and take up the thread of the inquiry exactly where Mendel dropped it.

As was stated in a lecture to the Royal Horticultural Society in 1900 it is to De Vries, Correns, and Tschermak that we owe the simultaneous rediscovery, confirmation and extension of Mendel's work. References* are there given to the chief recent publications relating to the subject, of which the number is rapidly increasing.

The whole paper abounds with matters for comment and criticism, which could only be profitable if undertaken at some length. There are also many deductions and lines of enquiry to which Mendel's facts point, which we in a fuller knowledge of physiology can perceive. It may, however, be doubted whether in his own day his conclusions could have been extended.

As some biographical particulars respecting this remarkable investigator will be welcome, I subjoin the following brief notice, which was published by Correns† on the authority of Dr. von Schanz: Gregor Johann Mendel was born on July 22, 1822, at Heinzendorf bei Odrau, in Austrian Silesia. He was the son of well-to-do peasants. In 1843 he entered as a novice the 'Koniginkloster', an Augustinian foundation in Altbrünn. In 1847 he was ordained priest. From 1851 to 1853 he studied physics and natural science at Vienna. Thence he returned to his cloister and became a teacher in the Realschule at Brünn. Subsequently he was made Abbot, and died January 6, 1884. The experiments described in his papers were carried out in the garden of his Convent.

Besides the two papers on hybridization, dealing respectively with *Pisum* and *Hieracium*, Mendel contributed to the Brünn journal observations of meteorological character, but, so far as I am aware, no others relating to natural history.—W. Bateson.

1 Introductory remarks

Artificial fertilization, such as is effected with ornamental plants in order to obtain new variations in colour, has led to the experiments which will be discussed here. The striking regularity with which the same hybrid forms always reappeared whenever fertilization took place between the same species induced further experiments to be undertaken, the object of which was to follow up the developments of the hybrids in their progeny.

---

* Abh. Naturf. Brünn, viii. 1869, p. 26.

* Journal Royal Horticultural Society, 1900, xxv. p. 54.

† Bot. Zeitg. lviii. 1900, No. 15, p. 229.

To this object numerous careful observers, such as Kölreuter, Gärtner, Herbert, Lecoq, Wichura and others, have devoted a part of their lives with tireless persistence. Gärtner especially, in his work 'Die Bastarderzeugung im Pflanzenreiche' (The Production of Hybrids in the Plant Kingdom), has recorded very valuable observations, and quite recently Wichura published the results of some profound investigations into the hybrids of the willow. That, so far, no generally applicable law governing the formation and development of hybrids has been successfully formulated can hardly be wondered at by anyone who is acquainted with the extent of the task, and can appreciate the difficulties with which experiments of this class have to contend. A final decision can only be arrived at when we have before us the results of detailed experiments made on plants belonging to the most diverse families.

Those who survey the work done in this field will arrive at the conviction that among all the numerous experiments made, not one has been carried out to such an extent and in such a way as to permit of the possibility of determining the number of different forms under which the offspring of hybrids appear, or so that these forms may be arranged with certainty according to their separate generations, or that their statistical relations can be definitely ascertained.

It requires indeed some courage to undertake a labour of such far-reaching extent; it appears, however, to be the only right way by which we can finally reach the solution of a question the importance of which cannot be overestimated in connection with the evolutionary history of organic forms.

The paper now presented records the results of such a detailed experiment. This experiment was appropriately confined to a small plant group, and is now, after 8 years' pursuit, concluded in all essentials. Whether the plan upon which the separate experiments were conducted and carried out was the best suited to attain the desired end is left to the friendly decision of the reader.

2 Selection of the trial plants

The value and validity of any experiment are determined by the fitness of the material to the purpose for which it is used, and thus in the case before us it cannot be immaterial what plants are subjected to experiment and in what manner such experiments are conducted.

The selection of the plant group which shall serve for experiments of this type must be made with all possible care if it be desired to avoid at the outset every risk of questionable results.

The experimental plants must necessarily:

1 Possess constant differing traits.

2 The hybrids of such plants must, during the flowering period, be protected from the influence of all foreign pollen, or be easily capable of such protection.

The hybrids and their offspring should suffer no marked disturbance in their fertility in the successive generations.

Contamination by foreign pollen, if such occurred during the experiments and were not recognized, would lead to entirely erroneous conclusions. Reduced fertility or entire sterility of certain forms, such as occurs in the offspring of many hybrids, would render the trials very difficult or entirely frustrate them. In order to discover the relations in which the hybrid forms stand towards each other and also towards their parental types it appears to be necessary that all members of the series occurring in each successive generation should be, *without exception*, subjected to observation.

At the very outset special attention was devoted to the Leguminosae on account of their peculiar floral structure. Experiments which were made with several members of this family led to the result that the genus *Pisum* was found to satisfy the necessary conditions.

Some thoroughly distinct forms of this genus possess traits which are constant and easily and reliably distinguishable, and yield perfectly fertile hybrid offspring from reciprocal crosses. Furthermore, interference by foreign pollen cannot easily occur, since the fertilizing organs are closely packed within the keel and the anther bursts within the bud, so that the stigma becomes covered with pollen even before the flower opens. This circumstance is of special importance. As additional advantages worth mentioning, there may be cited the easy culture of these plants in the open ground and in pots, and also their relatively short period of growth. Artificial fertilization is certainly a somewhat elaborate process, but nearly always succeeds. For this purpose the bud is opened before it is perfectly developed, the keel is removed, and each stamen carefully extracted by means of forceps, after which the stigma can be dusted at once with the foreign pollen.

In all, 34 more or less distinct varieties of peas were obtained from several seedsmen and subjected to a two years' trial. In the case of one variety among a large number of similar plants were noticed several which deviated considerably. These, however, did not vary in the following year, and agreed entirely with another variety obtained from the same seedsmen; doubtless the seeds had been accidentally mixed. All the other varieties yielded perfectly constant and similar offspring; at any rate, no essential difference was observed during the two trial years. For fertilization 22 of these were selected and planted annually during the whole period of the experiments. They remained constant without any exception.

Their systematic classification is difficult and uncertain. If we adopt the strictest definition of a species, according to which only those individuals belong to a species which under precisely the same circumstances display precisely similar traits, no two of them could be imputed to one species. According to the opinion of experts, however, the majority belong to the species *Pisum sativum*, while the rest are regarded and classed, some as sub-species of *P. sativum* and some as independent species, such as *P. quadratum*, *P. saccharantum*, and *P. umbellatum*. The positions, however, which may be assigned to them in a classificatory system are quite immaterial for the purposes of the experiments in question. It has so far been found to be just as impossible to draw a sharp line between the hybrids of species and varieties as between species and varieties themselves.

3 Division and arrangement of the experiments

When two plants which differ constantly in one or several traits are crossed, numerous experiments have demonstrated that the common traits are transmitted unchanged to the hybrids and their progeny; but each pair of differing traits, on the other hand, unite in the hybrid to form a new trait, which in the progeny of the hybrid is usually subject to changes. The object of the trial was to observe these variations in the case of each pair of differing traits, and to deduce the law according to which they appear in the successive generations. The trial resolves itself therefore into just as many separate experiments as there are constantly differing traits presented in the experimental plants.

The various forms of peas selected for crossing showed differences in the length and colour of the stem; in the size and shape of the leaves; in the position, colour and size of the flowers; in the length of the flower stalk; in the colour, shape and size of the pods; in the shape and size of the seeds; in the colour of the seed-coats and the cotyledons. Some of the traits noted do not permit of a sharp and certain separation, since the difference is of a 'more or less' nature, which is often difficult to define. Such traits could not be used for the separate experiments; these could only be confined to traits which stand out clearly and definitely in the plants. Lastly, the result should show whether they observe a concordant behaviour in their hybrid unions, and whether from these facts any conclusion can be arrived at regarding those traits which possess a subordinate significance in the classification.

The traits which were selected for the trials relate to:

1 *The difference in the shape of the ripe seeds.* These are either round or roundish, with depressions (when these occur on the surface) always very shallow, or they are irregularly angular and deeply wrinkled (*P. quadratum*).

2 *The difference in the colour of the seed cotyledon* (endosperm)*. The cotyledon of the ripe seeds is either pale yellow, bright yellow and orange coloured, or it possesses a more or less intense green tint. This difference of colour is easily seen in the seeds, as their coats are transparent.

3 *The difference in the colour of the seed-coat.* This is either white, in which case it is always associated with white flowers, or it is grey, grey-brown, leather-brown, with or without violet spotting, in which case the colour of the standards is violet, that of the wings purple, and the stem at the leaf axils is of a reddish tint. The grey seed-coats become black-brown in boiling water.

4 *The difference in the shape of the ripe pods.* These are either smoothly arched and in no way constricted, or they are deeply constricted between the seeds and more or less wrinkled (*P. saccharatum*).

5. *The difference in the colour of the unripe pods.* They are either light to dark green, or vividly yellow, in which colouring the stalks, leaf-veins, and calyx participate†.

6 *The difference in the position of the flowers.* They are either axial, that is, distributed along the main stem, or they are terminal, that is, bunched at the top of the stem and arranged almost in a short cyme (false umbel); in this case the upper part of the stem is more or less enlarged in cross-section (*P. umbellatum*).

7 *The difference in the length of the stem.*‡ The length of the stem varies a lot in individual varieties; it is, however, a constant trait for each, in so far that in healthy plants, grown in the same soil, it is only subject to unimportant variations.

In experiments with this trait, in order to be able to discriminate with certainty, the long axis of 6–7 ft was always crossed with the short one of $\frac{3}{4}$ ft to $1\frac{1}{2}$ ft.

Each two of the differing traits enumerated above as pairs were united by fertilization. For the:

1st experiment	60 fertilizations on 15 plants were undertaken
2nd experiment	58 fertilizations on 10 plants were undertaken
3rd experiment	35 fertilizations on 10 plants were undertaken
4th experiment	40 fertilizations on 10 plants were undertaken
5th experiment	23 fertilizations on 5 plants were undertaken
6th experiment	34 fertilizations on 10 plants were undertaken
7th experiment	37 fertilizations on 10 plants were undertaken

From a larger number of plants of the same variety only the most vigorous were chosen for fertilization. Weakly plants always afford uncertain results, because even in the first generation of hybrids, and still more so in the subsequent ones, many of the offspring either entirely fail to flower or form only a few inferior seeds.

* Mendel uses the terms 'albumen' and 'endosperm' somewhat loosely to denote the cotyledons, containing food material, within the seed. W.B.

† One species possesses a beautifully brownish-red coloured pod, which when ripening turns to violet and blue. Experiments with this trait were only begun last year. [Of these further experiments it seems no account was published. W.B.]

‡ In my account of these experiments (*RHS Journal*, vol. xxv, p. 54) I misunderstood this paragraph and took 'axis' to mean the *floral* axis, instead of the main axis of the plant. The unit of measurement, being indicated in the original by a dash, I thus took to have been an *inch*, but the translation here given is evidently correct. W.B.

Furthermore, in all the experiments, reciprocal crossings were effected in such a way that each of the two varieties which in one set of fertilizations served as seed bearers in the other set were used as pollen plants.

The plants were grown in garden beds, a few also in pots, and were maintained in their natural upright position by means of sticks, branches of trees and strings stretched between. For each experiment a number of pot plants were placed during the flowering period in a greenhouse, to serve as control plants for the main experiment in the open as regards possible attack by insects. Among the insects* which visit peas the beetle *Bruchus pisi* might be detrimental to the experiments should they appear in numbers. The female of this species is known to lay her eggs in the flower, and in so doing opens the keel; upon the tarsi of one specimen, which was caught in a flower, some pollen grains could be seen clearly with a lens. Mention must also be made of a circumstance which possibly might lead to the introduction of foreign pollen. It occurs, for instance, in some rare cases that certain parts of an otherwise quite normally developed flower are stunted, which results in a partial exposure of the fertilizing organs. Thus defective development of the keel was observed, owing to which the stigma and anthers remained partially uncovered. It also sometimes happens that the pollen does not reach full maturity. In this event there occurs a gradual lengthening of the stigma during the flowering period, until the tip of the stigma protrudes from the tip of the keel. This remarkable phenomenon has also been observed in hybrids of *Phaseolus* and *Lathyrus*.

The risk of adulteration by foreign pollen is, however, a very slight one with *Pisum*, and is quite incapable of affecting the general result. Among more than 10 000 plants which were carefully examined there were only a few cases where doubtless contamination had occurred. Since in the greenhouse such a case was never remarked, it may well be supposed that *Bruchus pisi*, and possibly also the described abnormalities in the floral structure, were to blame.

4 The forms of the hybrids†

Experiments which in previous years were made with ornamental plants have already afforded evidence that the hybrids, as a rule, are not exactly intermediate between the parental species. With some of the more striking characters (those, for instance, which relate to the form and size of the leaves, the pubescence of the several parts, etc.) the intermediate, indeed, was nearly always to be seen; in other cases, however, one of the two parental traits was so preponderant that it was difficult, or quite impossible, to detect the other in the hybrid.

This is precisely the case with pea hybrids. In the case of each of the seven crosses the hybrid trait resembles that of one of the parental forms so closely that the other either escapes observation completely or cannot be detected with certainty. This circumstance is of great importance in the determination and classification of the forms under which the offspring of the hybrids appear. Henceforth in this paper those traits which are transmitted entirely, or almost unchanged in the hybridization, and therefore in themselves represent the hybrid traits, are termed the *dominant*, and those which become latent in the process *recessive*. The expression 'recessive' has been chosen because the traits thereby designated withdraw or entirely disappear in the hybrids, but nevertheless reappear unchanged in their progeny, as will be demonstrated later on.

* It is somewhat surprising that no mention is made of thrips, which swarm in pea flowers. W.B.

† Mendel throughout speaks of his cross-bred peas as 'hybrids', a term which many restrict to the offspring of two distinct *species*. He, as he explains, held this to be only a question of degree. W.B.

It was furthermore shown by the whole of the experiments that it is completely immaterial whether the dominant trait belongs to the seed-bearer or to the pollen parent; the form of the hybrid remains identical in both cases. This interesting fact was also emphasized by Gärtner, with the remark that even the most practised expert is not in a position to determine in a hybrid which of the two parental species was the seed or the pollen plant.

Of the differentiating characters which were used in the experiments the following are dominant:

1 The round or roundish form of the seed with or without shallow depressions.

2 The yellow colouring of the seed cotyledons.

3 The grey, grey-brown, or leather-brown colour of the seed-coat, in connection with violet-red blossoms and reddish spots in the leaf axils.

4 The smoothly arched shape of the pod.

5 The green colouring of the unripe pod in connection with the same colour in the stems, the leaf-veins and the calyx.

6 The distribution of the flowers along the stem.

7 The length of the longer stem.

With regard to this last trait it must be stated that the longer of the two parental stems is usually exceeded by the hybrid, which is possibly only attributable to the greater luxuriance which appears in all parts of plants when stems of very different length are crossed. Thus, for instance, in repeated experiments, stems of 1 ft and 6 ft in length yielded without exception hybrids which varied in length between 6 ft and 7½ ft.

Hybrid seed coats are often more spotted, and the spots sometimes coalesce into small bluish-violet patches. The spotting also frequently appears even when it is absent as a parental trait.

The hybrid forms of the seed-shape and of the cotyledon colour are developed immediately after the artificial fertilization merely by the influence of the foreign pollen. They can, therefore, be observed even in the first trial year, while all the other traits naturally only appear in the following year in such plants as have been raised from the crossed seed.

5 The first generation from the hybrids

In this generation there reappear, together with the dominant traits, also the recessive ones with their individualities, and this occurs in the definitely expressed average proportion of three to one, so that among each four plants of this generation three receive the dominant trait and one the recessive. This relates without exception to all the traits which were embraced in the experiments. The angular wrinkled shape of the seed, the green colour of the cotyledon, the white colour of the seed-coats and the flowers, the constrictions of the pods, the yellow colour of the unripe pod, of the stalk, the calyx and the leaf veins, the almost umbellate inflorescence and the dwarfed stem, all reappear in the numerical proportion given without any essential alteration. *Transitional forms were not observed in any experiment.*

Since the hybrids resulting from reciprocal crosses were of identical appearance and presented no appreciable difference in their subsequent development, the results of both crosses can be reckoned together in each experiment. The relative numbers which were obtained for each pair of different traits are as follows:

Experiment 1 Shape of seed. From 253 hybrids 7 324 seeds were obtained in the second experimental year. Among them were 5 474 round or roundish ones and 1 850 angular wrinkled ones. This gives a ratio of 2.96 to 1.

Experiment 2 Colour of cotyledon. 258 plants yielded 8 023 seeds, 6 022 yellow, and 2 001 green; the ratio, therefore, is 3.01 to 1.

In these two experiments each pod yielded usually both kinds of seed. In well-developed pods which contained on average six to nine seeds, it often happened that all the seeds were round (Experiment 1) or all yellow (Experiment 2); on the other hand, never more than five angular or five green seeds were observed in one pod. It appears to make no difference whether the pods are developed early or later in the hybrid or whether they spring from the main stem or from an axillary one. In some few plants only a few seeds developed in the first-formed pods and these possessed exclusively one of the two traits, but in the subsequently developed pods the normal proportions were maintained nevertheless.

As in separate pods, so did the distribution of the traits vary in separate plants. By way of illustration, the first 10 individuals from both series of experiments may serve.

	Experiment 1 Shape of seed		Experiment 2 Colour of cotyledon	
Plants	Round	Angular	Yellow	Green
1	45	12	25	11
2	27	8	32	7
3	24	7	14	5
4	19	10	70	27
5	32	11	24	13
6	26	6	20	6
7	88	24	32	13
8	22	10	44	9
9	28	6	50	14
10	25	7	44	18

As extremes in the distribution of the two seed traits in one plant, there were observed in Experiment 1 an instance of 43 round and only 2 angular, and another of 14 round and 15 angular seeds. In Experiment 2 there was a case of 32 yellow and only 1 green seed, but also one of 20 yellow and 19 green.

These two experiments are important for the determination of the average relative figures, because with a smaller number of experimental plants they show that very considerable fluctuations may occur. In counting the seeds, also, especially in Experiment 2, some care is needed, since in some of the seeds of many plants the green colour of the cotyledon is less developed, and at first may be easily overlooked. The cause of the partial disappearance of the green colouring has no connection with the hybrid character of the plants, as it likewise occurs in the parental variety. This peculiarity is also confined to the individual and is not inherited by the offspring. In luxuriant plants this appearance was frequently noted. Seeds which are damaged by insects during their development often vary in colour and form, but with a little practice in sorting errors are easily avoided. It is almost superfluous to mention that the pods must remain on the plants until they are thoroughly ripened and have become dried, since it is only then that the shape and colour of the seed are fully developed.

Experiment 3 Colour of the seed-coats. Among 929 plants 705 bore violet-red flowers and grey-brown seed-coats; 224 had white flowers and white seed-coats. This gives a ratio of 3.15 to 1.

Experiment 4 Shape of pods. Of 1 181 plants 882 had smoothly arched pods and in 299 they were constricted. Ratio: 2.95 to 1.

Experiment 5 Colour of the unripe pods. The number of experimental plants was 580, of which 428 had green pods and 152 yellow ones. Ratio: 2.82 to 1.

Experiment 6 Position of flowers. Among 858 cases 651 blossoms were axial and 207 terminal. Ratio: 3.14 to 1.

Experiment 7 Length of stem. Out of 1 064 plants, in 787 cases the stem was long, and in 277 short. This gives a ratio of 2.84 to 1. In this experiment the dwarfed plants were carefully lifted and transferred to a special bed. This precaution was necessary, as otherwise their growth would have been stunted among their taller relatives. Even in their quite young state they can be picked out by their compact growth and thick dark-green foliage.

If now the results of all the experiments be brought together, there is found, as between the number of forms with the dominant and recessive traits, a ratio of 2.98 to 1, or 3 to 1.

The dominant trait can have here a *double significance*, that is, that of the parental characteristic, or the trait of the hybrid. In which of the two meanings it appears in each separate case can only be determined by the following generation. As a parental trait it must be transmitted unchanged to the whole of the offspring; as a hybrid trait, on the other hand, it must maintain the same behaviour as in the first generation.

6   The second generation from the hybrids

Those forms which in the first generation maintain the recessive trait do not further vary in the second generation as regards this trait; they remain constant in their offspring.

It is otherwise with those which possess the dominant trait in the first generation. Of these *two*-thirds yield offspring which display the dominant and recessive traits in the proportion of 3 to 1, and thereby show exactly the same ratio as the hybrid forms, while only *one*-third remains with the dominant trait constant.

The separate experiments yielded the following results:

Experiment 1   Among 565 plants which were raised from round seeds of the first generation, 193 yielded round seeds only, and remained therefore constant in this trait; 372, however, gave both round and angular seeds, in the proportion of 3 to 1. The number of the hybrids, therefore, as compared with the constants is 1.93 to 1.

Experiment 2   Of 519 plants which were raised from seeds whose cotyledon was of yellow colour in the first generation, 166 yielded exclusively yellow, while 353, however, yielded yellow and green seeds in the proportion of 3 to 1. There resulted, therefore, a splitting into hybrid and constant forms in the ratio 2.13 to 1.

For each separate experiment in the following experiments 100 plants were selected which displayed the dominant trait in the first generation, and in order to ascertain the significance of this, 10 seeds of each were cultivated.

Experiment 3   The offspring of 36 plants yielded exclusively grey-brown seed-coats, while 64 plants yielded some grey-brown and some white.

Experiment 4   The offspring of 29 plants had only smoothly arched pods; of the offspring of 71, on the other hand, some had smoothly arched and some constricted pods.

Experiment 5   The offspring of 40 plants had only green pods; of the offspring of 60 plants some had green, some yellow ones.

Experiment 6   The offspring of 33 plants had only axial flowers; of the offspring of 67, on the other hand, some had axial and some terminal flowers.

Experiment 7   The offspring of 28 plants inherited the long stem, and those of 72 plants some the long and some the short stem.

In each of these experiments a certain number of the plants came constant with the dominant trait. For the determination of the proportion in which the segregation of the forms with the constantly persistent trait results, the two first experiments are of especial importance, since in these a larger number of plants can be compared. The ratios 1.93 to 1 and 2.13 to 1 give together almost exactly the average ratio of 2 to 1. Experiment 6 has a quite concordant result; in the others the ratio varies more or less, as was only to be expected in view of the smaller number of 100 experimental plants. Experiment 5 which shows the greatest departure, was repeated, and instead of the ratio of 60 and 10 that of 65 and 35 resulted. *The average ratio of 2 to 1 appears, therefore, as fixed with certainty.* It is therefore demonstrated that, of those forms which possess the dominant trait in the first generation, in two-thirds the hybrid character is embodied, while one-third remains constant with the dominant trait.

The ratio of 3 to 1, in accordance with which the distribution of the dominant and recessive traits results in the first generation, resolves itself therefore in all experiments into the ratio of $2:1:1$ if one differentiates between the meaning of a dominant trait as a hybrid trait and as a parental characteristic. Since the members of the first generation spring directly from the seed of the hybrids, *it is now clear that the hybrids form seeds having one or other of the two differing traits, and of these one-half develop again the hybrid form, while the other half yield plants which remain constant and receive the dominant and recessive traits in equal numbers.*

7   The subsequent generations from the hybrids

The proportions in which the descendants of the hybrids develop and split up in the first and second generations presumably hold good for all subsequent progeny. Experiments 1 and 2 have already been carried through six generations, 3 and 7 through five, and 4, 5 and 6 through four, these experiments being continued from the third generation with a small number of plants, and no departure from the rule has been perceptible. The offspring of the hybrids separated in each generation in the ratio of $2:1:1$ into hybrids and constant forms.

If A be taken as denoting one of the two constant traits (for instance the dominant) with a for the recessive, and Aa for the hybrid form in which both are conjoined, the formula

$$A + 2Aa + a$$

shows the order of development for the progeny of the hybrids of two differing traits.

The observations made by Gärtner, Kölreuter, and others, that hybrids are inclined to revert to the parental forms, is also confirmed by the experiments described. It is seen that the number of hybrids which arise from one fertilization, as compared with the number of forms which become constant and the progeny of such from generation to generation, is continuously diminishing, but that nevertheless they could not entirely disappear. If there be assumed an average equality of fertility in all plants in all generations, and that, furthermore, each hybrid forms seed of which one-half yields hybrids again, while the other half is constant to both traits in equal proportions, the ratio of numbers for the offspring in each generation is seen by the following summary, in which A and a denote again the two parental traits, and Aa the hybrid forms. For brevity's sake it may be assumed that each plant in each generation furnishes only four seeds.

Generation	A	Aa	a	Ratios A : Aa : a		
1	1	2	1	1 :	2 :	1
2	6	4	6	3 :	2 :	3
3	28	8	28	7 :	2 :	7
4	120	16	120	15 :	2 :	15
5	496	32	496	31 :	2 :	31
n				$2^n - 1$:	2 :	$2^n - 1$

In the tenth generation, for instance, $2^n - 1 = 1\,023$. There result, therefore, in each 2 048 plants which arise in this generation, 1 023 with the constant dominant trait, 1 023 with the recessive trait, and only two hybrids.

8   The offspring of hybrids in which several differing traits are associated

In the experiments described above plants were used which differed only in one essential trait. The next task consisted in ascertaining whether the law of development discovered in these applied to each pair of differing traits when several diverse characters are united in the hybrid by crossing. As regards the form of the hybrids in these cases, the experiments showed throughout that this invariably more nearly approaches to that one of the two parental plants which possesses the greater number of dominant traits. If, for instance, the seed plant has a short stem, terminal white flowers, and smoothly arched pods, and the pollen plant, on the other hand, a long stem, violet-red flowers distributed along the stem, and constricted pods, the hybrid resembles the seed parent only in the form of the pod; in the other traits it agrees with the pollen parent. Should one of the two parental types possess only dominant traits, then the hybrid is scarcely or not at all distinguishable from it.

Two experiments were made with a greater number of plants. In the first experiment the parental plants differed in the shape of the seed and in the colour of the cotyledon; in the second in the form of the seed, in the colour of the cotyledon, and in the colour of the seed-coats. Experiments with seed traits give the result in the simplest and most certain way.

In order to facilitate study of the data in these experiments, the different characters of the seed plant will be indicated by A, B, C, those of the pollen plant by a, b, c, and the hybrid forms of these characters by Aa, Bb and Cc.

Experiment 1 AB, seed parents; ab, pollen parents;
A, shape round; a, shape angular;
B, cotyledon yellow. b, cotyledon green.

The fertilized seeds appeared round and yellow like those of the seed parents. The plants raised from these seeds yielded seeds of four types, which frequently presented themselves in one pod. In all, 556 seeds were yielded by 15 plants, and of these there were:

315 round and yellow,
101 angular and yellow,
108 round and green,
32 angular and green.

All were sown the following year. Eleven of the round yellow seeds did not germinate and three plants did not mature. Among the rest:

38 had round yellow seeds .. .. AB
65 round yellow and green seeds .. .. .. ABb
60 round yellow and angular yellow seeds .. .. AaB
138 round yellow and green, angular yellow and green
seeds .. .. AaBb

From the angular yellow seeds 96 resulting plants bore seed, of which:

28 had only angular yellow seeds aB
68 angular yellow and green seeds aBb

From 108 round green seeds 102 resulting plants fruited, of which:

35 had only round green seeds Ab
67 round and angular green seeds Aab

The angular green seeds yielded 30 plants which bore identical seeds; they remained constant.

The offspring of the hybrids appeared therefore under nine different forms, some in very unequal numbers. When these are summarized and co-ordinated we find:

38 plants with the designation AB
35 plants with the designation Ab
28 plants with the designation aB
30 plants with the designation ab
65 plants with the designation ABb
68 plants with the designation aBb
60 plants with the designation AaB
67 plants with the designation Aab
138 plants with the designation AaBb

The whole of the forms may be classed into three essentially different groups. The first embraces those with the signs AB, Ab, aB and ab; they possess only constant traits and do not vary again in the next generation. Each of these forms is represented on average 33 times. The second group embraces the signs ABb, aBb, AaB, Aab; these are constant in one trait and hybrid in another and vary in the next generation only as regards the hybrid trait. Each of these appears on average 65 times. The form AaBb occurs

138 times; it is hybrid in both traits, and behaves exactly as do the hybrids from which it is derived.

If the numbers in which the forms of these sections appear are compared, the ratios of 1, 2, 4 are unmistakably evident. The numbers 32, 65, 138 present very favourable approximations to the ratio numbers of 33, 66, 132.

The developmental series consists, therefore, of nine classes, of which four appear therein always once and are constant in both traits; the forms AB, ab resemble the parental forms, the two others represent the other possible constant combinations between the associated traits A, a, B, b. Four classes appear always twice, and are constant in one trait and hybrid in the other. One class appears four times, and is hybrid in both traits. Consequently the offspring of the hybrids, if two kinds of differing traits are combined therein, are developed according to the formula:

$$AB + Ab + aB + ab + 2ABb + 2aBb$$
$$+ 2ÁaB + 2Aab + 4AaBb.$$

This developmental series is incontestably a combination series in which the two developmental series for the traits A and a, B and b, are combined. We arrive at the full number of the classes of the series by the combination of the formulae:

$$A + 2Aa + a$$
$$B + 2Bb + b$$

Experiment 2
ABC, seed parents; abc, pollen parents;
A, shape round; a, shape angular;
B, cotyledon yellow; b, cotyledon green;
C, seed-coat grey-brown. c, seed-coat white.

This experiment was made in precisely the same way as the previous one. Among all the experiments it demanded the most time and trouble. From 24 hybrids 687 seeds were obtained in all: these were all either spotted, grey-brown or grey-green, round or angular. From these in the following year 639 plants fruited, and, as further investigation showed, there were among them:

8 plants ABC	22 plants ABCc	45 plants ABbCc
14 plants ABc	17 plants AbCc	36 plants aBbCc
9 plants AbC	25 plants aBCc	38 plants AaBCc
11 plants Abc	20 plants abCc	40 plants AabCc
8 plants aBC	15 plants ABBc	49 plants AabBC
10 plants aBc	18 plants ABbc	48 plants AaBbc
10 plants abC	19 plants aBbC	
7 plants abc	24 plants aBbc	
	14 plants AaBC	78 plants AaBbCc
	18 plants AaBc	
	20 plants AabC	
	16 plants Aabc	

The developmental series embraced 27 classes. Of these 8 are constant in all traits, and each appears on average 10 times; 12 are constant in two traits, and hybrid in the third, and each appears on average 19 times; 6 are constant in one trait and hybrid in the other two, and each appears on average 43 times. One form appears 78 times and is hybrid in all of the traits. The ratios 10, 19, 43, 78 agree so closely with the ratios 10, 20, 40, 80, or 1, 2, 4, 8, that this last undoubtedly represents the true value.

The development of the hybrids when the original parents differ in three traits results therefore according to the following formula:

$$ABC + ABc + AbC + Abc + aBC + aBc + abC$$
$$+ abc + 2ABCc + 2AbCc + 2aBCc + 2abCc + 2ABbC$$
$$+ 2ABbc + 2aBbC + 2aBbc + 2AaBC + 2AaBc + 2AabC$$
$$+ 2Aabc + 4ABbCc + 4aBbCc + 4AaBCc + 4AabCc$$
$$+ 4AaBbC + 4AaBbc + 8AaBbCc.$$

Here also is involved a combination series in which the developmental series for the traits A and a, B and b, C and c, are united.

The formulae:

$$A + 2Aa + a$$
$$B + 2Bb + b$$
$$C + 2Cc + c$$

give all the classes of the series. The constant combinations which occur therein agree with all combinations which are possible between the traits A, B, C, a, b, c; two of them, ABC and abc, resemble the two original parental stocks.

In addition further experiments were made with a smaller number of experimental plants in which the remaining traits by twos and threes were united as hybrids; all yielded approximately the same results. There is therefore no doubt that for the whole of the traits involved in the experiments the principle applies that *the offspring of the hybrids in which several essentially different traits are combined represent the components of a series of combinations, in which the developmental series for each two different traits are associated.* It is demonstrated at the same time that *the relation of each two different traits in hybrid connection is independent of the other differences in the two original parental stocks.*

If *n* represents the number of characteristic differences in the two original stocks, 3^n gives the number of components of the combination series, 4^n the number of individuals which belong to the series, and 2^n the number of combinations which remain constant. The series therefore embraces, if the original stocks differ in four traits, $3^4 = 81$ component classes, $4^4 = 256$ individuals, and $2^4 = 16$ constant forms; stated differently, among each 256 offspring of the hybrids there are 81 different combinations, 16 of which are constant.

All constant combinations which in peas are possible by the combination of the said seven characteristic features were actually obtained by repeated crossing. Their number is given by $2^7 = 128$. Thereby is simultaneously given the practical proof *that the constant traits which appear in various forms of a plant group may be obtained in all the associations which are possible according to the laws of combination by means of repeated artificial fertilization.*

Experiments to determine the flowering time of the hybrids are not yet completed. It can, however, already be stated that the period stands almost exactly between those of the seed and pollen parents, and that the development of the hybrids with respect to this trait probably happens in the same way as in the case of the other traits. The forms which are selected for experiments of this nature must have a difference of at least 20 days from the mean date of blooming; furthermore, the seeds when sown must all be placed at the same depth in the soil, so that they may germinate simultaneously. Also, during the whole flowering period, the more important variations in temperature must be taken into account, and the partial hastening or delaying of the flowering which may result from this. It is clear this experiment presents many difficulties to be overcome and necessitates great attention.

If we endeavour to collate in a brief form the results arrived at, we find that those differing traits which admit of easy and certain recognition in the experimental plants all behave exactly alike in their hybrid associations. The offspring of the hybrids of each pair of differing traits are: one-half, hybrid again; the remainder, constant in equal proportions with the traits of the seed and pollen parents respectively. If several differing traits are combined by cross-fertilization in a hybrid, the resulting offspring form the components of a combination series in which the developmental series for each pair of differing traits are united.

The uniformity of behaviour shown by the whole of the characteristics submitted to experiment permits, and fully justifies, the acceptance of the principle that a similar relation exists in the other traits which appear less sharply defined in plants, and therefore could not be included in the separate experiments. An experiment with flower stems of different lengths gave on the whole a fairly satisfactory result, although the differentiation and serial arrangement of the forms could not be effected with that certainty which is indispensable for correct experiment.

9 The reproductive cells of hybrids

The results of the previously described experiments induced further experiments, the results of which appear fitted to afford some conclusions as regards the composition of the germinal and pollen cells of hybrids. An important basis for argument is afforded in *Pisum* by the circumstance that among the progeny of the hybrids constant forms appear, and that this occurs, too, in all combinations of the associated traits. So far as experience goes, we find it confirmed in every case that constant progeny can only be formed when the germinal cells and the fertilizing pollen are of like character, so that both are provided with the material for creating quite similar individuals, as is the case with the normal fertilization of pure strains. We must therefore regard it as essential that exactly similar factors are also at work in the production of the constant forms in the hybrid plants. Since the various constant forms are produced in *one* plant, or even in *one* flower of a plant, the conclusion appears logical that in the ovaries of the hybrids there are formed as many sorts of germinal cells, and in the anthers as many sorts of pollen cells, as there are possible constant combination forms, and that these germinal and pollen cells agree in their internal composition with those of the separate forms.

In point of fact it is possible to demonstrate theoretically that this hypothesis would fully suffice to account for the development of the hybrids in the separate generations, if we might at the same time assume that the various kinds of germinal and pollen cells were formed in the hybrids on average in equal numbers.

In order to prove these assumptions experimentally, the following experiments were selected. Two forms which were constantly different in the form of the seed and the colour of the cotyledon were united by fertilization.

If the differing traits are again indicated as A, B, a, b, we have:

AB, seed parent;	ab, pollen parent;
A, shape round;	a, shape angular;
B, cotyledon yellow.	b, cotyledon green.

The artificially fertilized seeds were sown together with several seeds of both original stocks, and the most vigorous examples were chosen for the reciprocal crossing. There were fertilized:

1 The hybrids with the pollen of AB
2 The hybrids with the pollen of ab
3 AB with the pollen of the hybrids
4 ab with the pollen of the hybrids.

For each of these four experiments all of the flowers on three plants were fertilized. If the above theory be correct, there must be developed on the hybrids germinal and pollen cells of the forms AB, Ab, aB, ab, and there would be combined:

1 The germinal cells AB, Ab, aB, ab with the pollen cells AB
2 The germinal cells AB, Ab, aB, ab with the pollen cells ab
3 The germinal cells AB with the pollen cells AB, Ab, aB, ab
4 The germinal cells ab with the pollen cells AB, Ab, aB, ab.

From each of these trials there could then result only the following forms:

1 AB, ABb, AaB, AaBb
2 AaBb, Aab, aBb, ab
3 AB, ABb, AaB, AaBb
4 AaBb, Aab, aBb, ab.

If, furthermore, the several forms of the germinal and pollen cells of the hybrids were produced on average in equal numbers, then in each experiment the said four combinations should stand in equal ratio to each other. A perfect agreement in the numerical relations was, however, not to be expected, since in each fertilization, even in normal cases, some germinal cells remain undeveloped or subsequently die, and many even of the well-formed seeds fail to germinate when sown. The above assumption is also limited to the extent that, while it demands the formation of an equal number of the various sorts of germinal and pollen cells, it does not require that this should apply to each separate hybrid with mathematical exactness.

The first and second experiments had pre-eminently the object of proving the composition of the hybrid germinal cells, while the third and fourth experiments were to decide that of the pollen cells. As is shown by the above demonstration, the first and second experiments and the third and fourth experiments should produce precisely the same combinations, and even in the second year the result should be partially visible in the form and colour of the artificially fertilized seed. In the first and third experiments the dominant traits of shape and colour, A and B, appear in each union, and are also partly constant and partly in hybrid union with the recessive traits a and b, for which reason they must impress their peculiarity upon the whole of the seeds. All seeds should therefore appear round and yellow, if the theory be justified. In the second and fourth experiments, on the other hand, one union is hybrid in shape and in colour, and consequently the seeds are round and yellow; another is hybrid in form, but constant in the recessive trait of colour, whence the seeds are round and green; the third is constant in the recessive trait of shape but hybrid in colour and consequently the seeds are angular and yellow; the fourth is constant in both recessive traits, so that the seeds are angular and green. In both these experiments there were consequently four sorts of seed to be expected: round yellow, round green, angular yellow, angular green.

The crop fulfilled these expectations perfectly. There were obtained in the:

1st experiment 98 exclusively round yellow seeds;

3rd experiment 94 exclusively round yellow seeds.

In the second experiment, 31 round yellow, 26 round green, 27 angular yellow, 26 angular green seeds.

In the fourth experiment 24 round yellow, 25 round green, 22 angular yellow, 27 angular green seeds.

A favourable result could now scarcely be doubted; the next generation must afford the final proof. From the seed sown there resulted for the first experiment 90 plants, and for the third 87 plants which fruited; these yielded for the:

1st experiment	3rd experiment		
20	25	round yellow seeds ..	AB
23	19	round yellow and green seeds	ABb
25	22	round and angular yellow	AaB
22	21	round and angular green and yellow seeds ..	AaBb

In the second and fourth experiments the round yellow seeds yielded plants with round and angular, yellow and green seeds, AaBb.

From the round green seeds, plants resulted with round and angular green seeds, Aab.

The angular yellow seeds gave plants with angular yellow and green seeds, aBb.

From the angular green seeds, plants were raised which yielded again only angular green seeds, ab.

Although in these two experiments also some seeds did not germinate, the figures arrived at already in the previous year were not affected thereby, since each kind of seed gave plants which, as regards their seed, were like each other and different from the others. There resulted therefore from the:

2nd experiment	4th experiment	
31	24	plants of the form AaBb
26	25	plants of the form Aab
27	22	plants of the form aBb
26	27	plants of the form ab

In all the experiments, therefore, there appeared all the forms which the proposed theory demands, and also in nearly equal numbers.

In a further experiment the traits of flower colour and length of stem were experimented upon, and selection so made that in the third experimental year each trait ought to appear in half of all the plants if the above theory were correct. A, B, a, b serve again as indicating the different traits.

A, violet-red flowers,   a, white flowers.

B, stem long,   b, stem short.

The form Ab was fertilized with ab, which produced the hybrid Aab. Furthermore, aB was also fertilized with ab, whence the hybrid aBb. In the second year, for further fertilization, the hybrid Aab was used as seed parent, and hybrid aBb as pollen parent.

seed parent, Aab,   pollen parent, aBb.

possible germinal cells, Ab, ab,   pollen cells, aB, ab.

From the fertilization between the possible germinal and pollen cells four combinations should result:

$$AaBb + aBb + Aab + ab$$

From this it is perceived that, according to the above theory, in the third experimental year out of all the plants:

The half should have violet-red flowers (Aa), Classes 1, 3

The half should have white flowers (a), Classes 2, 4

The half should have a long stem (Bb), Classes 1, 2

The half should have a short stem (b), Classes 3, 4.

From 45 fertilizations of the second year 187 seeds resulted, of which only 166 reached the flowering stage in the third year. Among these the separate classes appeared in the numbers following:

Class	Colour of flower	Stem	
1	violet-red	long	47 times
2	white	long	40 times
3	violet-red	short	38 times
4	white	short	41 times.

There consequently appeared:

The violet-red flower colour (Aa) in 85 plants

The white flower colour (a) in 81 plants

The long stem (Bb) in 87 plants

The short stem (b) in 79 plants.

The theory adduced is therefore satisfactorily confirmed in this experiment also.

For the traits of shape of pod, colour of pod, and position of flowers, experiments were also made on a small scale, and results obtained in perfect agreement. All combinations which were possible through the union of the differing traits duly appeared, and in nearly equal numbers.

Experimentally, therefore, the theory is justified that *pea hybrids form germinal and pollen cells which, in their constitution, represent in equal numbers all constant forms which result from the combination of the traits when united by fertilization.*

The difference of the forms among the progeny of the hybrids, as well as the ratios in which they are observed, find a sufficient explanation in the principle above deduced.

The simplest case is afforded by the developmental series for one pair of differing traits. This series is expressed by the formula A + 2Aa + a, in which A and a signify the forms with constant differing traits, and Aa the hybrid form of both. It includes in three different classes four individuals. In the formation of these, pollen and germinal cells of the form A and a on average take part equally in the fertilization; hence each form twice, since four individuals are formed. There participate consequently in the fertilization:

The pollen cells   A + A + a + a

The germinal cells A + A + a + a.

It remains, therefore, purely a matter of chance which of the two sorts of pollen will become united with each separate germinal cell. According, however, to the law of probability, it will always happen, on the average of many cases, that each pollen form A and a will unite equally often with each germinal cell form A and a. Consequently one of the two pollen cells A in the fertilization will meet with the germinal cell A and the other with a germinal cell a, and so likewise one pollen cell a will unite with a germinal cell A, and the other with a germinal cell a.

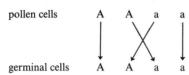

The result of the fertilization may be made clear by putting the signs for the associated germinal and pollen cells in the form of fractions, those for the pollen cells above and those for the germinal cells below the line. We then have:

$$\frac{A}{A} + \frac{A}{a} + \frac{a}{A} + \frac{a}{a}$$

In the first and fourth factor the germinal and pollen cells are of like kind. Consequently the product of their union must be constant, that is, A and a; in the second and third, on the other hand, there again results a union of the two differing traits of the stocks. Consequently the forms resulting from these fertilizations are identical with those of the hybrid from which they sprang. *There occurs accordingly a repeated hybridization.* This explains the striking fact that the hybrids are able to produce, besides the two parental forms, offspring which are like themselves; (A/a) and (a/A) both give the same union Aa, since, as already remarked above, it makes no difference in the result of fertilization to which of the two traits the pollen or germinal cells belong. We may write then:

$$\frac{A}{A} + \frac{A}{a} + \frac{a}{A} + \frac{a}{a} = A + 2Aa + a$$

This represents the average result of the self-fertilization of the hybrids when two differing traits are associated in them. In solitary flowers and in solitary plants, however, the ratios in which the members of the series are produced may suffer considerable fluctuations. Apart from the fact that the numbers in which both sorts of germinal cells occur in the ovary can only be regarded as equal on the average, it remains purely a matter of chance which of the two sorts of pollen may fertilize each separate germinal cell. For this reason the separate values must necessarily be subject to fluctuations, and there are even extreme cases possible, as were described earlier in connection with the experiments with the shape of the seed and the colour of the cotyledon. The true ratios of the numbers can only be ascertained by an average deduced from the sum of as many single values as possible; the greater the number the more are merely chance elements eliminated.

The developmental series for hybrids in which two kinds of differing traits are associated contains among 16 individuals 9 different forms, AB + Ab + aB + ab + 2ABb + 2aBb + 2AaB + 2Aab + 4AaBb. Between the differing traits of the original stocks Aa and Bb, 4 constant combinations are possible, and consequently the hybrids produce the corresponding 4 forms of the germinal and pollen cells AB, Ab, aB, ab, and each of these will on average either fertilize or be fertilized, since 16 individuals are included in the series. Therefore, the participators in the fertilization are:

pollen cells AB + AB + AB + AB + Ab + Ab + Ab + Ab
  + aB + aB + aB + aB + ab + ab + ab + ab

germinal AB + AB + AB + AB + Ab + Ab + Ab + Ab
cells + aB + aB + aB + aB + ab + ab + ab + ab

In the process of fertilization each pollen form unites on average equally often with each germinal cell form, so that each of the four pollen cells AB unites once with one of the forms of germinal cell AB, Ab, aB, ab. In precisely the same way the rest of the pollen cells of the forms Ab, aB, ab unite with all the other germinal cells.

We obtain therefore:

$$\frac{AB}{AB} + \frac{AB}{Ab} + \frac{AB}{aB} + \frac{AB}{ab} + \frac{Ab}{AB} + \frac{Ab}{Ab} + \frac{Ab}{aB} + \frac{Ab}{ab} +$$
$$\frac{aB}{AB} + \frac{aB}{Ab} + \frac{aB}{aB} + \frac{aB}{ab} + \frac{ab}{AB} + \frac{ab}{Ab} + \frac{ab}{aB} + \frac{ab}{ab}$$

or

AB + ABb + AaB + AaBb + ABb + Ab + AaBb + Aab
+ AaB + AaBb + aB + aBb + AaBb + Aab + aBb + ab
  = AB + Ab + aB + ab + 2ABb + 2aBb + 2AaB + 2Aab
  + 4AaBb*

In precisely similar fashion is the developmental series of hybrids exhibited when three kinds of differing traits are combined in them. The hybrids form eight various kinds of germinal and pollen cells—ABC, ABc, AbC, Abc, aBC, aBc, abC, abc—and each pollen form unites itself again on average once with each form of germinal cell.

The law of combination of different traits which governs the development of the hybrids finds therefore its foundation and explanation in the proven statement, that the hybrids produce germinal cells and pollen cells which in equal numbers represent all constant forms which result from the combination of traits united by fertilization.

10 Experiments with hybrids of other species of plants

It must be the object of further experiment to ascertain whether the law of development discovered for *Pisum* applies also to the hybrids of other plants. To this end several experiments were recently commenced. Two minor experiments with species of *Phaseolus* have been completed, and may be here mentioned.

An experiment with *Phaseolus vulgaris* and *Phaseolus nanus* gave results in perfect agreement. *Ph. nanus* had together with the dwarf stem smoothly arched green pods. *Ph. vulgaris* had, on the other hand, a stem 10 feet to 12 feet high, and yellow coloured pods, constricted when ripe. The ratios of the numbers in which the different forms appeared in the separate generations were the same as with *Pisum*. Also the development of the constant combinations resulted according to the law of simple combination of traits, exactly as in the case of *Pisum*. There were obtained:

Constant combinations	Stem	Colour of the unripe pods	Form of the ripe pods
1	long	green	arched
2	long	green	constricted
3	long	yellow	arched
4	long	yellow	constricted
5	short	green	arched
6	short	green	constricted
7	short	yellow	arched
8	short	yellow	constricted

The green colour of the pod, the arched pod shape, and the long stem were, as in *Pisum*, dominant traits.

Another experiment with two very different species of *Phaseolus* had only a partial result. *Phaseolus nanus*, L., served as seed parent, a perfectly constant species, with white flowers in short bunches and small white seeds in straight arched smooth pods; as pollen parent was used *Ph. multiflorus*, W., with tall winding stem, purple-red flowers in very long bunches, rough sickle-shaped crooked pods, and large seeds which bore black flecks and splashes on a peach-blossom-red ground.

The hybrids had the greatest similarity to the pollen parent, but the flowers appeared less intensely coloured. Their fertility was very limited; from 17 plants, which together developed many hundreds

---

* In the original the sign of equality (=) is here represented by +, evidently a misprint. W.B.

of flowers, only 49 seeds in all were obtained. These were of medium size, and were flecked and splashed similarly to those of *Ph. multiflorus*, while the ground colour was not materially different. The next year 44 plants were raised from these seeds, of which only 31 reached the flowering stage. The traits of *Ph. nanus*, which had been altogether latent in the hybrids, reappeared in various combinations; their ratio, however, with relation to the dominant traits was necessarily very fluctuating owing to the small number of experimental plants. With certain traits as in those of the stem and the shape of pod, it was, however, as in the case of *Pisum*, almost exactly 1 : 3.

Insignificant as the results of this experiment may be as regards the determination of the relative numbers in which the various forms appeared, it presents, on the other hand, the phenomenon of a remarkable change of colour in the flowers and seeds of the hybrids. In *Pisum* it is known that the traits of the flower- and seed-colour present themselves unchanged in the first and further generations, and that the offspring of the hybrids display exclusively the one or the other of the traits of the original stocks. It is otherwise in the experiment we are considering. The white flowers and the seed-colour of *Ph. nanus* appeared, it is true, at once in the first generation in one fairly fertile example, but the remaining 30 plants developed flower colours which were of various grades from purple-red to pale violet. The colouring of the seed pod was no less varied than that of the flowers. No plant could rank as fully fertile; many produced no fruit at all; others only yielded fruits from the flowers last produced, and did not ripen. Only from 15 plants were well-developed seeds obtained. The greatest disposition to infertility was seen in the forms with preponderantly red flowers, since out of 16 only 4 yielded ripe seed. Three of these had a similar seed pattern to *Ph. multiflorus*, but with a more or less pale ground colour; the fourth plant yielded only one seed of plain brown tint. The forms with preponderantly violet coloured flowers had dark brown, black-brown and quite black seeds.

The experiment was continued through two more generations under similar unfavourable circumstances, since even among the offspring of fairly fertile plants there were still some which were less fertile or even quite sterile. Other flower- and seed-colours than those cited did not subsequently present themselves. The forms which in the first generation contained one or more of the recessive traits remained, as regards these, constant without exception. Also of those plants which possessed violet flowers and brown or black seed, some did not vary again in these respects in the next generation; the majority, however, yielded, together with offspring exactly like themselves, some which displayed white flowers and white seed-pods. The red-flowering plants remained so poor that nothing can be said with certainty as regards their further development.

Despite the many disturbing factors with which the observations had to contend, it is nevertheless seen by this experiment that the development of the hybrids, with regard to those traits which concern the form of the plants, follows the same laws as does *Pisum*. With regard to the colour traits, it certainly appears difficult to perceive a substantial agreement. Apart from the fact that from the union of a white and a purple-red colouring a whole series of colours results, from purple to pale violet and white, the circumstance is a striking one that among 31 flowering plants only one received the recessive trait of the white colour, while in *Pisum* this occurs on the average in every fourth plant.

Even these enigmatical results, however, might probably be explained by the law governing *Pisum* if we might assume that the colour of the flowers and seeds of *Ph. multiflorus* is a combination of two or more entirely independent colours, which individually act like any other constant trait in the plant. If the flower colour A were a combination of the individual traits $A_1 + A_2 + \cdots$ which produce the total impression of a purple coloration, then by fertilization with the differing traits of the white colour, a, there would be produced the hybrid unions $A_1a + A_2a + \cdots$ and so would it be with the corresponding colouring of the seed-pods. According to the above assumption, each of these hybrid colour unions would be independent, and would consequently develop

quite independently from the others. It is then easily seen that from the combination of the separate developmental series a perfect colour-series must result. If, for instance, $A = A_1 + A_2$ then the hybrids A_1a and A_2a form the developmental series

$$A_1 + 2A_1a + a$$
$$A_2 + 2A_2a + a$$

The members of this series can enter into nine different combinations, and each of these denotes another colour

1 A_1A_2	2 A_1aA_2	1 A_2a
2 A_1A_2a	4 A_1aA_2a	2 A_2aa
1 A_1a	2 A_1aa	1 aa

The figures prescribed for the separate combinations also indicate how many plants with the corresponding colouring belong to the series. Since the total is 16, the whole of the colours are on average distributed over each of the 16 plants, but, as the series itself indicates, in unequal proportions.

Should the colour development really happen in this way, we could offer an explanation of the case above described, that the white flowers and seed-pod colour only appeared once among 31 plants of the first generation. This colouring appears only once in the series, and could therefore also only be developed once on average in each 16, and with three colour traits only once even in 64 plants.

It must, however, not be forgotten that the explanation here attempted is based on a mere hypothesis, only supported by the very imperfect result of the experiment just described. It would, however, be well worth while to follow up the development of colour in hybrids by similar experiments, since it is probable that in this way we might learn the significance of the extraordinary variety in the colouring of our ornamental flowers.

So far, little at present is known with certainty beyond the fact that the colour of the flowers in most ornamental plants is an extremely variable trait. The opinion has often been expressed that the stability of the species is greatly disturbed or entirely upset by cultivation, and consequently there is an inclination to regard the development of cultivated forms as a matter of chance devoid of rules; the colouring of ornamental plants is indeed usually cited as a model of instability. It is, however, not clear why the simple transference into garden soil should result in such a thorough and persistent revolution in the plant organism. No one will seriously maintain that the development of plants in the open country is ruled by other laws than in the garden bed. Here, as there, changes of type must take place if the conditions of life be altered, and the species possesses the capacity to adapt itself to its new environment. It is willingly granted that by cultivation the origination of new varieties is favoured, and that by man's labour many varieties are acquired which, under natural conditions, would be lost; but nothing justifies the assumption that the tendency to the formation of varieties is so extraordinarily increased that species speedily lose all stability, and their offspring diverge into an infinite number of extremely variable forms. Were the change in the conditions of vegetation the sole cause of variability we might expect that those cultivated plants which are grown for centuries under almost identical conditions would again attain constancy. That, as is well known, is not the case, since it is precisely under such circumstances that not only the most varied but also the most variable forms are found. It is only the Leguminosae, like *Pisum*, *Phaseolus*, *Lens*, whose organs of fertilization are protected by the keel, which constitute a noteworthy exception. Even here there have arisen numerous varieties during a cultural period of more than 1 000 years under the most diverse conditions; these maintain, however, in unchanging environments a stability as great as that of species growing wild.

It is more than probable that as regards the variability of cultivated plants there exists a factor which so far has received little attention. Various experiments force us to the conclusion that our cultivated plants, with few exceptions, are *members of different hybrid series*, whose further development along regular lines is changed and hindered by frequent intraspecific crossings. The circumstance must not be overlooked that cultivated plants are mostly grown in great numbers and close together, which affords the most favourable conditions for reciprocal fertilization between the varieties present and the species itself. The probability of this is supported by the fact that among the great array of variable forms solitary examples are always found, which in one trait or another remain constant, if only foreign influence be carefully excluded. These forms develop precisely as do those which are known to be members of the compound hybrid series. Also with the most sensitive of all traits, that of colour, it cannot escape the careful observer that in the separate forms the inclination to vary is displayed in very different degrees. Among plants which arise from *one* spontaneous fertilization there are often some whose offspring vary widely in the constitution and arrangement of the colours, while others furnish forms of little deviation, and among a greater number solitary examples occur which transmit the colour of the flowers unchanged to their offspring. The cultivated species of *Dianthus* afford an instructive example of this. A white-flowered example of *Dianthus caryophyllus*, which itself was derived from a white-flowered variety, was shut up during its blooming period in a greenhouse; the numerous seeds obtained yielded plants entirely white-flowered like itself. A similar result was obtained from a subspecies, with red flowers somewhat flushed with violet, and one with flowers of white striped with red. Many others, on the other hand, which were similarly protected, yielded progeny which were more or less variously coloured and marked.

Whoever studies the coloration which results in ornamental plants from similar fertilization can hardly escape the conviction that here also the development follows a definite law which possibly finds its expression *in the combination of several independent colour traits.*

11 Concluding remarks

It can hardly fail to be of interest to compare the observations made regarding *Pisum* with the results arrived at by the two authorities in this branch of knowledge, Kölreuter and Gärtner, in their investigations. According to the opinion of both, the hybrids in external appearance present either a form intermediate between the original species, or they closely resemble either the one or the other type, and sometimes can hardly be distinguished from it. From their seeds usually arise, if the fertilization was effected by their own pollen, various forms which differ from the normal type. As a rule, the majority of individuals obtained by one fertilization maintain the hybrid form, while some few others come more like the seed parent, and an occasional individual approaches the pollen parent. This, however, is not the case with all hybrids without exception. With some, the offspring have more nearly approached, some the one and some the other, the original stock, or they all incline more to one or the other side; while with others *they remain perfectly like the hybrid* and continue constant in their offspring. The hybrids of varieties behave like hybrids of species, but they possess greater variability of form and a more pronounced tendency to revert to the original type.

With regard to the features of the hybrids and their development, as a rule an agreement with the observations made in *Pisum* is unmistakable. It is otherwise with the exceptional cases cited. Gärtner confesses even that the exact determination whether a form bears a greater resemblance to one or to the other of the two original species often involved great difficulty, so much depending upon the subjective point of view of the observer. Another circumstance could, however, contribute to render the results fluctuating and uncertain, despite the most careful observation and differentiation; for the experiments, plants were mostly used which rank as good species and are differentiated by a large number of traits. In addition to the sharply defined traits, where it is a question of greater or less similarity, those traits must also be taken into account which are often difficult to define in words, but yet suffice, as every plant connoisseur knows, to give such forms the appearance of a stranger. If it be accepted that the development of hybrids follows the law which is valid for *Pisum*, the series in each separate experiment must

embrace very many forms, since the number of the components, as is known, increases with the number of the differing traits as a power of three. With a relatively small number of experimental plants the result therefore could only be approximately right, and in single cases might fluctuate considerably. If, for instance, the 2 original stocks differ in 7 characters, and 100 and 200 plants were raised from the seeds of their hybrids to determine the grade of relationship of the offspring, we can easily see how uncertain the decision must become, since for 7 differing traits the developmental series contains 16 384 individuals under 2 187 various forms; now one and then another relationship could assert its predominance, just according as chance presented this or that form to the observer in a majority of instances.

If, furthermore, there appear among the differing traits at the same time dominant traits, which are transferred entire or nearly unchanged to the hybrids, then in the components of the developmental series that one of the two original stocks which possesses the majority of dominant traits must always be predominant. In the experiment described relative to *Pisum*, in which three kinds of differing traits were concerned, all the dominant traits belonged to the seed parent. Although the components of the series in their internal make-up approach both original stock plants equally, in this experiment the type of the seed parent obtained so great a preponderance that out of each 64 plants of the first generation 54 exactly resembled it, or only differed in one trait. It is seen how rash it may be under such circumstances to draw from the external resemblances of hybrids conclusions as to their internal relations.

Gärtner mentions that, in those cases where the development was regular among the offspring of the hybrids, the two parental types were not reproduced but only a few closely approximating individuals. With very extended developmental series it could not in fact be otherwise. For 7 differing traits, for instance, among more than 16 000 individuals (offspring of the hybrids) each of the 2 parental types would occur only once. It is hardly possible, therefore, that such should appear at all among a small number of experimental plants; with some probability, however, we might reckon upon the appearance of a few forms which approach them in the series.

We meet with an essential difference in those hybrids which remain constant in their progeny and propagate themselves as truly as the pure species. According to Gärtner, to this class belong the remarkably fertile hybrids *Aquilegia atropurpurea canadensis, Lavatera pseudolbia thuringiaca, Geum urbano-rivale* and some *Dianthus* hybrids, and, according to Wichura, the hybrids of the willow species. For the history of the evolution of plants this circumstance is of special importance, since constant hybrids acquire the status of new species. The correctness of this is evidenced by most excellent observers and cannot be doubted. Gärtner had opportunity to follow up *Dianthus armeria deltoides* to the tenth generation, since it regularly propagated itself in the garden.

With *Pisum* it was proved experimentally that the hybrids form germinal and pollen cells of different kinds, and that herein lies the reason of the variability of their offspring. In other hybrids, likewise, whose offspring behave similarly, we may assume a like cause; for those, on the other hand, which remain constant the assumption appears justifiable that their fertilizing cells are all alike and agree with the primordial-cell of the hybrid. In the opinion of renowned physiologists, for the purpose of propagation one pollen cell and one germinal cell unite in phanerogams* into a single cell, which is capable by assimilation and formation of new cells to develop an

---

* In *Pisum* it is placed beyond doubt that for the formation of the new embryo a perfect union of the elements of both fertilizing cells must take place. How could we otherwise explain that among the offspring of the hybrids both original types reappear in equal numbers and with all their peculiarities? If the influence of the germinal cell upon the pollen cell were only external, if it fulfilled the role of a foster mother only, then the result of each artificial fertilization could be no other than that the developed hybrid should exactly resemble the pollen parent, or at any rate do so very closely. This the experiments so far have in no way confirmed. An

independent organism. This development follows a constant law, which is founded on the material composition and arrangement of the elements which meet in the cell in a vivifying union. If the reproductive cells be of the same kind and agree with the primordial cell of the mother plant, then the development of the new individual will follow the same law which rules the mother plant. If it chances that a germinal cell unites with a dissimilar pollen cell, we must then assume that between those elements of both cells, which determine the mutual differences, some sort of compromise is effected. The resulting compound cell becomes the foundation of the hybrid organism, the development of which necessarily follows a different law from that obtaining in each of the two original species. If the compromise be taken to be a complete one, in the sense, namely, that the hybrid embryo is formed from cells of like kind in which the differences are entirely and permanently accommodated together, the further result follows that the hybrids, like any other stable plant species, remain true to themselves in their offspring. The reproductive cells which are formed in their ovaries and anthers are of one kind, and agree with the compound cell from which they derive.

With regard to those hybrids whose progeny is variable we may perhaps assume that between the differing elements of the germinal and pollen cells there also occurs a compromise, to the extent that the formation of a cell that becomes the basis of the hybrid becomes possible; but, nevertheless, the arrangement between the conflicting elements is only temporary and does not endure throughout the life of the hybrid plant. Since in the habit of the plant no changes are perceptible during the whole period of vegetation, we must further assume that it is only possible for the differing elements to liberate themselves from the enforced union when the fertilizing cells are developed. In the formation of these cells all existing elements participate in an entirely free and equal arrangement, in which it is only the differing ones which mutually separate themselves. In this way the production would be rendered possible of as many sorts of germinal and pollen cells as there are combinations possible of the formative elements.

The attribution attempted here of the essential difference in the development of hybrids to *a permanent or temporary union* of the differing cell elements can, of course, only claim the value of an hypothesis for which the lack of definite data offers a wide field. Some justification of the opinion expressed lies in the evidence afforded by *Pisum* that the behaviour of each pair of differing traits in hybrid union is independent of the other differences between the two original plants and, further, that the hybrid produces just so many kinds of germinal and pollen cells as there are possible constant combination forms. The differing traits of two plants, however, can finally depend only upon differences in the composition and grouping of the elements which exist in the fundamental cells in dynamic interaction.

Even the validity of the law formulated for *Pisum* still requires to be confirmed, and a repetition of the more important experiments is consequently much to be desired, that, for instance, relating to the composition of the hybrid fertilizing cells. A distinguishing feature may easily escape the single observer, which at the outset may appear to be unimportant but may later attain such significance that it must not be ignored in the total result. Whether the variable hybrids of other plant species observe an entire agreement must also be first decided experimentally. In the meantime we may assume that in material points a difference in principle can scarcely occur, since the unity in the plan of development of organic life is beyond question.

In conclusion, the experiments carried out by Kölreuter, Gärtner, and others with respect to *the transformation of one species into another by artificial fertilization* merit special mention. A special importance has been attached to these experiments, and Gärtner reckons them among 'the most difficult of all in hybridization'.

---

evident proof of the complete union of the contents of both cells is afforded by the experience gained on all sides that it is immaterial, as regards the form of the hybrid, which of the original species is the seed parent or which the pollen parent.

Should a species A be transformed into a species B, both would be united by fertilization and the resulting hybrids then be fertilized with the pollen of B; then out of the various offspring resulting that form would be selected which stood in nearest relation to B and once more would be fertilized with B pollen, and so continuously until finally a form was arrived at which was like B and constant in its progeny. By this process the species A would change into the species B. Gärtner alone has effected 30 such trials with plants of genera *Aquilegia*, *Dianthus*, *Geum*, *Lavatera*, *Lychnis*, *Malva*, *Nicotiana* and *Oenothera*. The length of time required for transformation was not alike for all species. While with some a triple fertilization sufficed, with others this had to be repeated five or six times, and even in the same species fluctuations were observed in various experiments. Gärtner ascribes this difference to the circumstance that 'the typical force by which a species, during reproduction, effects the change and transformation of the maternal type varies considerably in different plants, and that, consequently, the periods must also vary within which the one species is changed into the other, as also the number of generations, so that the transformation in some species is perfected in more and in others in fewer generations'. Further, the same observer remarks 'that in these transformation experiments a good deal depends upon which type and which individuals be chosen for further transformation'.

If it may be assumed that in these experiments the development of the forms resulted in a similar way to that of *Pisum*, the entire process of transformation would find a fairly simple explanation. The hybrid forms as many kinds of germinal cells as there are constant combinations possible of the traits associated within, and one of these is always of the same kind as the fertilizing pollen cells. Consequently there always exists the possibility with all such experiments that even from the second fertilization there may result a constant form identical with that of the pollen parent. Whether this really be obtained depends in each separate case upon the number of the experimental plants, as well as upon the number of differing traits which are united by the fertilization. Let us, for instance, assume that the plants selected for experiment differed in three traits, and the species ABC is to be transformed into the other species abc by repeated fertilization with the pollen of the latter; the hybrids resulting from the first cross form eight different kinds of germinal cells:

$$ABC, ABc, AbC, aBC, Abc, aBc, abC, abc$$

These in the second experimental year are united again with the pollen cells abc, and we obtain the series:

$$AaBbCc + AaBbc + AabCc + aBbCc + Aabc + aBbc$$
$$+ abCc + abc$$

Since the form abc occurs once in the series of eight components, it is consequently unlikely that it would be missing among the experimental plants, even were these raised in a smaller number, and the transformation would be perfected already by a second fertilization. If by chance it did not appear, then the fertilization must be repeated with one of those forms nearest to it, Aabc, aBbc, abCc. It is perceived that such an experiment must last the longer *the smaller the number of experimental plants and the larger the number of differing traits* in the two original species, and that, furthermore, in the same species there can easily occur a delay of one or even of two generations such as Gärtner observed. The transformation of widely divergent species could generally only be completed in five or six experimental years, since the number of different germinal cells which are formed in the hybrid increases with the square of the number of differing traits.

Gärtner found by repeated experiments that the respective period of transformation varies in many species, so that frequently a species A can be transformed into a species B a generation sooner than can species B into species A. He deduces from this that Kölreuter's opinion can hardly be maintained that 'the two natures in hybrids are perfectly in equilibrium'. It appears, however, that Kölreuter does not merit this criticism, but rather that Gärtner has overlooked a material point, to which he himself elsewhere draws attention, i.e. that 'it depends which individual is chosen for further transformation'. Experiments which in this connection were carried out with two species of *Pisum* demonstrated that as regards the choice of the fittest individuals for the purpose of further fertilization it may make a great difference which of two species is transformed into the other. The two experimental plants differed in five traits, while at the same time those of species A were all dominant and those of species B all recessive. For mutual transformation A was fertilized with pollen of B, and B with pollen of A, and this was repeated with both hybrids the following year.

With the first experimental (B/A) there were 87 plants available in the third experimental year for the selections of individuals for further crossing, and these were of the possible 32 forms; with the second experimental (A/B) 73 plants resulted, which *with regard to their external appearance were identical with that of the pollen plant*; in their internal composition, however, they must have been just as varied as the forms of the other experiment; A definite selection was consequently only possible with the first experiment; with the second some plants selected at random were chosen. Of the latter only a proportion of the flowers were crossed with the A pollen, the others being left to fertilize themselves. Among the 5 plants which were selected in both experiments for fertilization agreement with the pollen parent was as follows:

First experiment	Second experiment	
2 plants	—	in all traits
3 plants	—	in 4 traits
—	2 plants	in 3 traits
—	2 plants	in 2 traits
—	1 plant	in 1 trait

In the first experiment, therefore, the transformation was completed; in the second, which was not continued further, two more fertilizations would probably have been required.

Although the case may not frequently occur that the dominant traits belong exclusively to one or the other of the original parent plants, it will always make a difference which of the two possesses the majority. If the pollen parent shows the majority, then the selection of forms for further crossing will afford a smaller degree of security than in the reverse case. A delay in the period of transformation, provided that the experiment is only considered as completed when a form is arrived at which not only exactly resembles the pollen plant in form but also remains as constant in its progeny, will result.

Gärtner, by the results of these transformation experiments, was led to oppose the opinion of those naturalists who dispute the stability of plant species and believe in a continuous evolution of plant forms. He perceives in the complete transformation of one species into another an indubitable proof that species are fixed within limits beyond which they cannot change. Although this opinion cannot be unconditionally accepted, we find on the other hand in Gärtner's experiments a noteworthy confirmation of the supposition regarding variability of cultivated plants which has already been expressed.

Among the experimental species there were cultivated plants, such as *Aquilegia atropurpurea* and *canadensis*, *Dianthus caryophyllus*, *chinensis* and *japonicus*, *Nicotiana rustica* and *paniculata*, and hybrids between these species lost none of their stability after four or five generations.

Self-assessment questions

SAQ 1 Given below (Fig. 13) is a pedigree for inheritance of an odd hair pattern in humans:

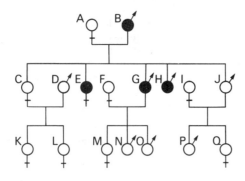

Figure 13

(a) How many children has J?

(b) Who is the eldest of B's children?

(c) Name the individuals showing the trait.

(d) How many daughters has F?

(e) What is the relationship of H to N?

SAQ 2 A woman, A, with normal hips marries a man, B, with a deformed hip. The first two children are both sons (C and D) and have normal hips. The daughter (E) has a deformed hip. The first-born son, C, marries a woman, F, with normal hips. They have three sons: the first-born (G) and the third son (I) are normal; the second son (H) has a deformed hip. Man, D, marries woman, J, but they have no children. Woman, E, marries a normal man (K). They have three daughters (L, M and N). Only the first-born (L) has a deformed hip; the second-born (M) and the youngest (N) are normal.

Represent this hypothetical example using the conventions developed in this Unit (refer to Fig. 3). Is the trait (deformed hip) dominant or recessive? Is there anything odd?

SAQ 3 In 1905, Bateson, Saunders and Punnett studied inheritance in the sweet pea (*Lathyrus odoratus*). They found that the flower colour, either purple or red, was due to a single gene. Likewise, pollen shape, long or round, also depended on alternative alleles of a single gene. They then investigated plants differing in both characteristics. A cross between a purple-flowered plant with long pollen and a red-flowered plant with round pollen yielded F_1 plants that all had purple flowers and long pollen. The F_2 plants were as follows:

purple flower, long pollen	284
purple flower, round pollen	21
red flower, long pollen	21
red flower, round pollen	55

Are these data consistent with Mendel's law of independent assortment?

Answers to entry test

1 False. They are held together by hydrogen bonds between the nucleotide bases.

2 False. The number remains the same.

3 True.

4 False. The sequence of bases determines the sequence of amino acids.

5 False. They contain half the number of chromosomes.

6 True.

7 False. It is semi-conservative.

8 True.

9 False. It arises by fusion of two gamete cells.

10 False. The code is a triplet one. The mRNA must therefore be at least 300 nucleotides long.

If you scored less than 7 out of 10 correct, then you are advised to check back in the appropriate Sections of S100. If you scored 7 out of 10 or more, continue with the Unit now.

Answers to ITQs

ITQ 1 (*Objective 1*) Though the data from Figure 5 is insufficient, it is typical of data from many such studies. It appears that: (a) not all members of a family (see the second line of the chart) carry either the non-anaemic trait or the disease; (b) the non-anaemic trait and disease occur in both sexes; (c) the trait occurs in two generations shown here (lines 2 and 3); (d) there appears to be no obvious correlation between the disease or the non-anaemic trait and the order of birth.

In addition, in the only instances of anaemics (individuals E and F), both parents (C and D) show the non-anaemic trait. The significance of this will become apparent later on.

ITQ 2 (*Objective 1*) The disease is manifest only in males.

ITQ 3 (*Objectives 1 and 5*) The ratio of dominant: recessive is:

$$\text{round : wrinkled} \quad \approx 2:9$$
$$\text{yellow : green} \quad \approx 3:0$$
$$\text{pigmented : non-pigmented} \approx 3:1$$

In each case, the ratio of dominant type: recessive type in the F_2 is close to $3:1$. Or, expressed another way, the dominant type is three-quarters of the total F_2 progeny, the recessive type one-quarter.

ITQ 4 (*Objectives 1 and 4*) yy green; YY yellow; Yy yellow.

ITQ 5 (*Objective 4*) Consider the true-breeding parents for green seeds. The genotype must be yy (see ITQ 4). The gametes it produces contain only y. An egg that contains y is fertilized by a pollen grain that contains y, to give a zygote containing a pair of alleles, yy. It is green in phenotype and will eventually, as an adult, yield only y type gametes, and so on. Likewise, for a true-breeding plant for yellowness.

49

ITQ 6 (*Objective 4*)

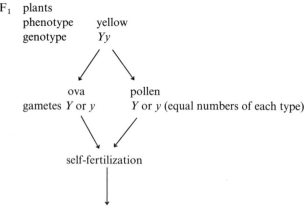

F_1 plants
   phenotype   yellow
   genotype Yy

                   ova   pollen
       gametes Y or y Y or y (equal numbers of each type)

                        self-fertilization

F_2 (a) Y egg $\times$ y pollen $=$ Yy yellow
   (b) Y egg $\times$ Y pollen $=$ YY yellow
   (c) y egg $\times$ Y pollen $=$ yY yellow
   (d) y egg $\times$ y pollen $=$ yy green

Therefore, the proportion of phenotype in the F_2 is three-quarters yellow, one-quarter green. That is, yellow:green in the F_2 is $3:1$. Note that this ratio also *depends on the occurrence of gametes of different genotypes in equal numbers.* Indeed, this was one of the central conclusions from Mendel's studies (Radio programme 1 and Mendel's paper).

The four possible combinations of gametes (a)–(d) is more readily symbolized by aligning the gametes along the sides of a 'two by two' square and 'multiplying' to obtain the zygote genotype inside the four smaller squares, as below:

	eggs	
	y	Y
y	yy	yY
Y	Yy	YY

pollen

ITQ 7 (*Objective 5*) (a) The pedigree is consistent with albinism's being due to a single gene. This will be obvious from the answers below (b)–(d). Let us label this gene, P, when in dominant allelic form and, p, when in the recessive form.

(b) The first generation at which albinos actually occur is in the third line (see Fig. 8). Two children of the mating between man A and woman B are albino. Phenotypes like albinism must originally occur via chance changes, so-called *mutations*, from the normal. Because such mutations are rare it is unlikely that the trait would arise in this way in two brothers (C and D). Therefore, it is much more likely that the two brothers (C and D) inherit the trait from one or other, or both, parents. Thus, the parents (A and/or B) probably carry the trait in some way.

(c) We have now established that one or both of A and B probably carry the trait. If, however, P meant albinism, then any individual carrying just one allele for that trait would be albino. A and B are not. *Therefore the allele for albinism must be recessive, p.* Only individuals homozygous for p are albino. Hence, P is for pigment, p for albino. Hence, A and B are probably both of genotype Pp. Their offspring can be normally pigmented (PP or Pp) or albino (pp).

(d) The father, C, is albino and hence of genotype pp. The mother is non-albino and therefore of genotype PP or Pp. If she was of genotype PP all her children would inherit P allele from her and therefore be Pp—normally pigmented. She must, therefore, be Pp—heterozygous with respect to the gene concerned with pigmentation.

ITQ 8 (*Objective 2*) The usual method of detection is to examine the *phenotype* not the *genotype* directly. Where dominance occurs, it will not always be easy to distinguish between organisms homozygous for the dominant allele and those heterozygous for it. In other words, the presence of the recessive allele may be completely masked in heterozygotes. This is awkward as it is genotype frequencies one is interested in calculating, not phenotype frequencies.

ITQ 9 (*Objective 5*) Assuming that both groups of blacks come from common ancestral populations, one would expect that, as the incidence of malaria has been low for several generations in the USA, the frequency of the Hb^S allele in the modern American black will have fallen to a lower level than in the modern African Negro whose habitat remains malarial.

Answers to SAQs

SAQ 1 (*Objective 4*) (a) 2; (b) C; (c) B, E, G, H; (d) 1; (e) H is N's uncle. If you were wrong on any of these questions you are advised to study again the conventions explained on p. 10.

SAQ 2 (*Objectives 4 and 5*) We have labelled the individuals in Figure 14 to enable you to check your answer. Normally we would not label them.

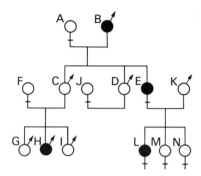

Figure 14

First, we shall ignore the unlikely possibility that a further mutation would lead to the occurence of the trait for a second time in the family in question. Now to the question of dominance. Man H shows the trait, yet both his parents (F and C) are normal. Thus, F and C must be heterozygous for the trait which is recessive. If we represent normal hips as due to allele P and abnormal as caused by p, then F and C are both Pp and H is pp. Likewise, L, E and B are pp. Therefore, K must be Pp as L is the daughter of E and K (pp) and is pp. Likewise, A must be Pp, as E is pp. F must also be Pp. The odd thing in that the allele p is in at least three apparently unconnected families: the main family (A and B), F and K. This is a rather high incidence of a rare trait (Unit 14).

SAQ 3 (*Objectives 3 and 5*) No. From the F_1 data, purple flowers and long pollen are the dominant phenotypes for the two traits being considered. Thus, if the F_2 of Mendel's law was obeyed, we would expect a ratio of the four types; purple–long, purple–round, red–long, red–round, of $9:3:3:1$ ($56\%:19\%:19\%:7\%$ approx). In fact, the ratio among the total 381 plants is $284/381:21/381:21/381:55/381$ or approximately $74\%:6\%:6\%:14\%$. This appears not to be consistent with the $56:19:19:6$ expected from Mendel's law. However, when you have listened to Radio programme 1 you might like to reconsider this SAQ and do a χ^2 test to check whether, in fact, the deviation from the expected data really makes this experiment inconsistent with Mendel's law. Further information bearing on this issue will come from Unit 3.

Acknowledgements

Grateful acknowledgement is made to the following for material used in this Unit:

Text

Robert Conquest, 'Common sense on colour blindness', *Daily Telegraph*, 20 April 1974.

Illustrations

Figure 1 National Institute of Health, Bethesda, Maryland; *Figure 2* Austrian National Library Picture Archive; *Figure 10* from Tsutomu Watanabe, 'Infectious drug resistance', *Scientific American*, December 1967; *Figure 12 Ogonyok*.

2   Chromosomes and Genes

S299
GENETICS

Contents

List of scientific terms used in Unit 2

Introduced in S100*	Developed in this Unit	Page No.
allele	autoradiography	65
chloroplast	autosome	87
DNA	autotrophic	102
dominant	auxotroph	82
endoplasmic reticulum	bivalent	78
gene	cell cycle	73
genotype	chiasmata	78
haploid	chromatid	70
heterozygote	chromatin	63
homozygote	colchicine	72
mathematical indices	complementation	98
(e.g. 10^6; 10^{-9}; etc.)	complementation map	99
mitochondria	conditional lethal mutants	90
nucleus	eukaryote	64
phenotype	Feulgen stain	60
recessive	genome	96
ribosome	heterokaryon	100
RNA	heterogametic	85
	heterotrophic	102
	homogametic	85
	homologous chromosomes	77
	independent assortment	84
	interphase	69
	lethal mutants	89
	meiosis	76
	mitosis	69
	mutagen	97
	mutation	88
	nucleoids	61
	phases—S, G_1, G_2	74
	prokaryote	64
	prototroph	82
	pulse-labelling	74
	segregation analysis	82
	semi-conservative replication	68
	semi-lethal mutants	89
	sex chromosomes	85
	sex linkage	87
	sex-linked lethal mutation	92
	synapsis	78
	temperature sensitivity	93
	transformation	58

* The Open University (1971) S100 *Science: A Foundation Course*, The Open University Press.

Objectives for Unit 2

After studying this Unit, you should be able to:

1 Define, recognize the best definition of, and place in the correct context, the items in the list of scientific terms opposite.

2 Describe the differences between prokaryotes and eukaryotes.
(ITQ 4; SAQ 1)

3 Describe the principal features of mitosis and meiosis and show the genetic significance of both processes using data on segregating genes.
(SAQs 2, 3 and 4)

4 Describe the replication and fidelity of genetic material at the nucleic acid and chromosomal levels.
(ITQ 8)

5 Recall the chemical composition of chromosomes and provide evidence from studies with radioactive precursor molecules of DNA for chromosome continuity through cell divisions and interphases.
(SAQs 2 and 3)

6 Using mammals and insects, specifically, show how sex differences are genotypically determined by specific chromosomes.
(SAQs 5 and 8)

7 Show that genetic analysis is possible only after the examination of progeny resulting from the fusion between the genetic material of different individuals of the same (or closely related) species.
(SAQs 5, 6 and 8)

8 Predict the phenotype and genotype of progeny from genetic crosses and human pedigrees for no more than two independently segregating genes.
(SAQs 5 and 8)

9 Explain how heritable variation (mutation) occurs as a result of changes in genes.
(SAQ 9)

10 Calculate and compare, from given data, the frequencies of mutation for different characteristics in different organisms.

11 Explain by reference to specific examples, the general principles of the induction of mutants by chemicals and radiation.
(SAQs 6, 7 and 10)

12 Describe the nature of complementation tests using heterozygotes, and explain how they are used to show that identity of phenotype does not necessarily imply identity of genotype.
(SAQ 9)

13 Carry out a simple statistical test (χ^2) on genetic data.
(ITQ 12)

Study guide for Unit 2

Unit 2 is the first of three Units (2, and 3 and 4) that provide you with many of the essential concepts and tools of genetics. It is most important that you have a thorough grasp of the contents of these Units because they are essential if you are to follow the arguments and ideas that are developed later in the Course. Specifically, Unit 2 introduces you to chromosomes and genes, and you will encounter both of these structures in virtually every Unit that follows. In Units 3 and 4, the concepts of linkage, the independent assortment of genes and recombination are considered. Later, you will see how the genetic analysis of fusion and segregation enables the geneticist to understand and predict the outcome of genetic crosses.

Inevitably there are 'loose ends' in these early Units. Many ideas mentioned only briefly at this stage will be taken up and elaborated on later in the Course. For example, you will discover in Unit 5 that there is a great deal to be added to your knowledge of the structure of chromosomes, of chromosome interrelationships and alterations to chromosomes (mutations). In Unit 6, which deals with molecular genetics, you will be encouraged to relate the structure of DNA to genetic analysis, as this in turn is an essential preliminary to looking at cytoplasmic inheritance (Unit 7) and the genetics of development (Unit 8).

For Units 7 and 8, it is also important that you have a clear idea of the distinction between prokaryotes and eukaryotes.

Indeed, when you have completed Unit 2 and Units 3 and 4, you should have acquired the basic skills to tackle the quantitative genetics of Units 9–13 inclusive and the human genetics (Units 14 and 15), in which you will find many of the concepts and information from earlier Units reiterated and used in different contexts. For example, in Units 9 and 10 we shall be looking at mutations again, but this time dealing with the prospect for the survival of new mutations (both advantageous and deleterious) in a population.

But, to return to Unit 2—the important points to bear in mind are:

1   DNA (deoxyribose nucleic acid) is the chemical basis of heredity in all living organisms, with the exception of a few viruses that have RNA (ribose nucleic acid) in place of DNA.

2   The organization of DNA into discrete structures called chromosomes is a feature of higher organisms, but not of bacteria where the DNA is 'naked' within the cell.

3   The replication of DNA and the duplication and separation of chromosomes are separate events preceding cell division that should be considered within the total context of the cell cycle.

4   Genes can be defined by segregation analysis, that is, analysis of the progeny following fusion between genetic material of different individuals of the same (or closely related) species.

5   In order to make genetic analysis possible, there have to be differences between alleles; these differences arise by mutation. In the TV programme for this Unit you will begin to see how the geneticist first finds and then verifies a mutant.

For many of you, topics such as mitosis and meiosis will already be familiar, and the Unit should provide useful revision of these subjects. As for the remainder of the Unit, our experience from developmental testing suggests that the material is not too difficult and that the whole Unit can be completed well within the allotted study time. Nevertheless, there are Sections in the text, especially those where radio-actively labelled molecules are described, that may be unfamiliar to you and on which you may have to spend a little more time, for example, the techniques employed in determining semi-conservative replication in DNA or pulse-labelling to elucidate aspects of the cell cycle. Again, if you are unsure about the idea of a chemical isotope you might like to check briefly with Units 6 and 7 of S100 before beginning Unit 2.

To fulfil Objective 13 you will need to be able to employ the goodness-of-fit χ^2 test of significance. By now you should have read up to Section ST.4 of the Statistics text* (see the study guide to *STATS*), which gives you the background you require.

Finally, we urge you to try the Entry Test on p. 57 before you begin on the main story, and to complete your study of Unit 2 by answering the SAQs on pp. 102–4.

* The Open University (1976) S299 STATS *Statistics for Genetics*, The Open University Press. This text is to be studied in parallel with the Units of the Course. From now on we shall refer to it by its code *STATS*.

Entry test for Unit 2

Which of the following statements are *true* and which are *false*?

1   DNA is degraded by DNAase.

2   RNA is the same as DNA except for the substitution of uracil for thymine as one of its four bases.

3   Bacteria are unicellular organisms.

4   An enzyme is a catalyst.

5   All enzymes are proteins.

6   A mixture of molecules can be separated by ultracentrifugation.

7   All radioactive isotopes can be detected with a Geiger–Müller counter.

8   Many unicellular organisms reproduce by binary fission.

9   Amino acids are joined together by mRNA to make proteins.

10   Bacteriophages are viruses that attack bacteria.

Label the numbered structures in Figure 1.

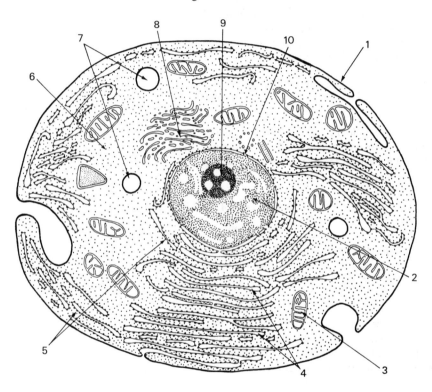

Figure 1 A generalized representation of a cell seen under the electron microscope.

Now check your answers against those given on p. 104. If you did not score at least 15 out of a possible 20 points, you should reread Units 14 and 17 of S100, particularly Sections 14.4, 17.1, 17.2, 17.3, 17.4, 17.5, 17.9 and 17.10.

2.0   Introduction to Unit 2

Unit 1 and TV programme 1 should have impressed upon you the importance and significance of Mendel's experiments. As a result of these, geneticists of the twentieth century have been able to capitalize on the idea of a factor (gene) to rationalize the results of breeding experiments. Later in this Unit, we shall look at the ways in which the algebra of formal genetic analysis enables us to explain some of the results of breeding experiments. The second landmark in the history of genetics was the confluence of genetic analysis (Mendelism) and cytology in the early part of this century. It resulted in the chromosome theory of heredity. The discovery of the vehicles of heredity (chromosomes) within cells and, eventually, the elucidation of their chemical composition (nucleic acids) has helped to make genetics the most powerful tool in the solution of virtually all biological problems.

In this Unit, we shall attempt to answer the following questions:

1   How can we locate genes in the cell?

2   What is the chemical composition of chromosomes?

3   Are the observed changes in the behaviour of chromosomes compatible with segregation phenomena?

4   How can we identify genes?

5   What are mutants and how frequently do they occur?

6   Do mutants of the same phenotype necessarily result from the same genetic changes?

Please keep these questions in mind while you are studying Unit 2 and refer back to this Introduction whenever necessary.

2.1   The theory that DNA is the chemical basis of heredity

Since the work of Miescher (1871) on the nuclei of pus cells, it has been known that the main constituent of the cell nucleus is nucleoprotein; a combination of basic* proteins and nucleic acids. Later, it was discovered that the genes of all organisms are composed of DNA (deoxyribose nucleic acid) with the exception of certain viruses whose genetic material is composed of RNA (ribose nucleic acid).

Although we assume that you know that genes are actually composed of DNA (S100[1]), let us briefly remind you of some of the indirect and direct lines of evidence supporting this theory.

First, if we examine the amounts of DNA per diploid nucleus in various somatic cell types of the same species, we find that they are remarkably similar. For example, Table 1 shows the mean values per diploid nucleus of DNA in the somatic cells of the chicken (values expressed in picograms (grams $\times$ 10^{-12})).

Table 1   The mean values per diploid nucleus of DNA in the somatic cells of the chicken (picograms)

Spleen	Red blood cell	Liver cell	Heart cell	Kidney cell	Pancreas
2.50	2.49	2.66	2.45	2.20	2.61

Considering the small values involved and the limitations of the chemical assay techniques, these values are remarkably constant. In contrast, values for the basic proteins associated with the nuclei of these different somatic cells are extremely variable. Nevertheless, this in itself does not constitute support for the theory that genes are composed of DNA.

> ITQ 1   Can you explain why this observation does not constitute support for the theory that genes are composed of DNA?

The answers to the ITQs are on pp. 105–8.

Direct evidence about whether protein or DNA is the genetic material came from the classical experiments of Avery, Macleod and McCarty in 1944 (see *HIST*†), which showed genetic transformation in a bacterium, *Diplococcus pneumoniae*. The idea of transforming one strain of bacterium to another was not new and was based on earlier work by Griffith in 1928. What Avery and his colleagues did was to indicate which chemical substance was primarily responsible for this *transformation*. They used two strains of pneumonia bacterium; one (RII) that does not have an external polysaccharide capsule to its cell membrane and another (SIII) that does. When

transformation

---

* The term 'basic' protein is used in contrast to 'acidic' protein.

† The Open University (1976) S299 HIST, *The History and Social Relations of Genetics*, The Open University Press. This text is to be studied in parallel with the Units of the Course. We refer to it by its code *HIST*.

grown on nutrient agar (a culture medium), the colonies (each made up of about a million bacteria derived by binary fission from an initial single bacterium) of RII have a roughened surface, but those of SIII are smooth. This difference in the surface structure of the two types of bacterial cell is responsible for the difference in appearance of the two types of colonies.

When both strains were mixed together in a test-tube and a sample of the mixture was spread on nutrient agar and incubated, colonies appeared as the bacteria multiplied (Fig. 2). (Note that the diagrams do not represent the actual number of colonies found.)

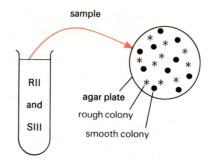

Figure 2

When the SIII strain was 'heat-killed' before it was mixed with the RII strain and a sample of the mixture was again grown on agar, the results were as shown in Figure 3.

QUESTION How do you account for the appearance of the smooth colonies (albeit very few)?

ANSWER *Either*: (i) not all the SIII strain bacteria were killed by heat, *or* (ii) some substance(s) from the dead SIII cells was (were) able to affect the characteristics of some of the RII cells, *or* (iii) spontaneous mutation from RII type to SIII type occurred.

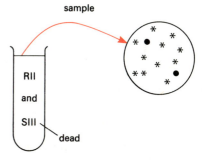

Figure 3

Hypothesis (i) was tested by spreading a pure sample of heat-killed SIII cells on nutrient agar and looking for colonies. None grew.

Hypothesis (ii) above was tested by mixing RII cells with various fractions that had been isolated from heat-killed SIII cells (i.e. a protein fraction, a carbohydrate fraction, a DNA fraction, an RNA fraction, etc.). The following results were obtained (Fig. 4).

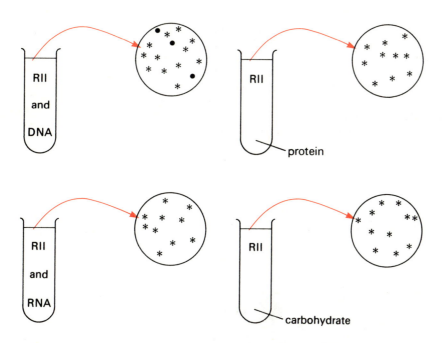

Figure 4

QUESTION What do you infer from these results?

ANSWER DNA is the substance that has affected the structure of the RII bacteria.

Hypothesis (iii) was tested by culturing RII bacteria alone over many generations. No smooth colonies were seen in any generation.

QUESTION What conclusions do you draw from this experiment?

ANSWER Spontaneous mutation to the SIII type is an unlikely explanation for the appearance of the smooth colonies.

QUESTION If DNA were the genetic material responsible for this transformation, what colonies would you expect from experiments in which:

(a) DNAase, an enzyme which degrades (destroys) DNA, was added to:
(i) RII plus SIII whole cells (previously heat-killed*) and (ii) RII plus DNA
extracted from heat-killed SIII;
(b) RNAase, which degrades RNA, was added to RII plus DNA extracted
from heat-killed SIII?

ANSWER   (a) (i) and (ii) Rough colonies only. (b) Rough and smooth colonies.

These predictions were found to be correct. Since these experiments were carried out,
the importance of DNA in heredity has been conclusively demonstrated by Hershey
and Chase with bacteriophage (see *HIST*) and, subsequently, DNA's identical role
in other organisms has been shown by many other workers.

2.2   The location of DNA in cells

You already know that genes are composed of DNA (or occasionally of RNA in
certain viruses), and that, in some cells at least, the likely location of DNA is the
nucleus.

ITQ 2   How could you confirm the specific location of DNA in cells?

The most reliable methods for locating DNA in the cell are those involving the use
of radioactive isotopes or the use of specific stains.

However, even staining DNA is not as simple as it may sound, because many basic
dyes react with a variety of cellular components and are not necessarily specific to
DNA. In fact, it was not until 1924 when Feulgen introduced the stain that now
bears his name, that the first specific stain for DNA was found. The stain will not
react with the ribose of RNA and hence it distinguishes DNA from RNA because
the amino groups ($-NH_2$) of the Feulgen compound will combine with the aldehyde   **Feulgen stain**
($-CHO$) groups produced after mild hydrolysis of the deoxyribose sugar of DNA.
The DNA-dye complex so formed is deep magenta in colour. Since Feulgen's
discovery, the technique has been refined and modified so that it can now be used as a
quantitative measure for DNA. Quantitative estimations are possible because the
concentration or density of the DNA-dye complex formed is directly related to the
amount of light transmitted through or absorbed by the specimen under consideration and thus measurements can be made with instruments such as a colorimeter
or a spectrophotometer.

From the light micrographs in Figure 5 the DNA appears to be located in two
densely staining areas in the cells of *Esherichia coli* (see the cell marked X) and within
the nucleus or nuclear region of the plant cells (*Vicia faba*).

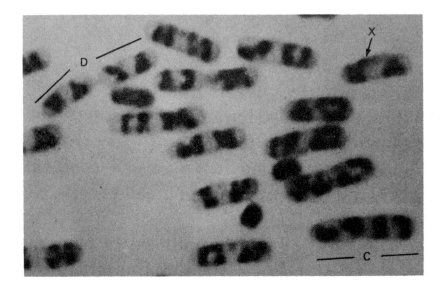

Figure 5(a)   Light micrograph of a
bacterium, *E. coli*, showing the location
of DNA by a staining technique.
($\times 3\,000$)

* Note that the critical temperature, selected for heat-killing, denatures the organism's
enzymes but does not seriously affect the nucleic acids.

Even when the bacterial cell is dividing (marked C and D in Fig. 5(a)), the stain is unable to resolve any further detail in structure within the stained areas. In contrast, when the cell is dividing, the nucleus and the chromosomes can be seen to take up the stain in preference to their surrounding cytoplasm (see Fig. 5(b)). Further, it can be shown that these chromosomes account for about 99 per cent of the total DNA in the cell.

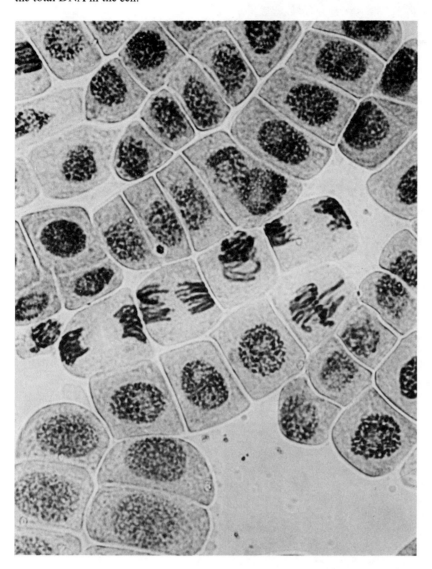

Figure 5(b) Light micrograph of the root-tip cells of the broad-bean, *Vicia Faber*, showing the location of DNA by a staining technique. (× 1 250)

The chromosomes (or 'coloured bodies') are a distinctive feature of eukaryotic cells but are not found in prokaryotic cells. As we shall be using the terms prokaryotic and eukaryotic intermittently throughout this and other Units, let us take the opportunity here to clarify them.

2.3 Prokaryotes and eukaryotes

QUESTION Examine the structures labelled N in the electron micrographs (Fig. 6(a), (b) and (c), *overleaf*) and in each case state whether a nuclear membrane is present or not.

ANSWER A nuclear membrane is absent in *E. coli* but present in both the plant and animal cells.

The two 'lighter' areas that seem to contain fine threads in the bacterial cell are not surrounded by a nuclear membrane and do not possess a nucleolus. Such areas in the bacterial cell are referred to as *nucleoids* or *chromatin bodies*, to distinguish them from the nuclei shown in the other two electron micrographs, in which nuclear membranes

nucleoids

61

(a)

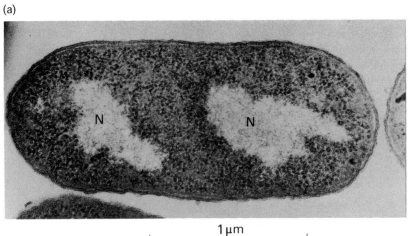

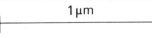

1 μm

(b)

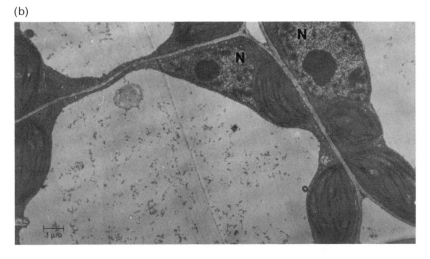

(c)

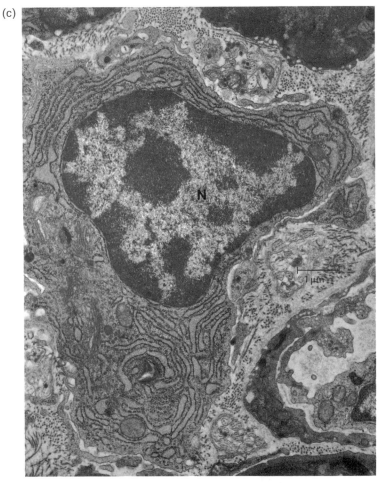

Figure 6 Three electron micrographs of different cells: (a) a bacterium, *E. coli*; (b) a cell from the leaf of a plant; (c) an animal cell, from the gut of a mouse. N—nucleus.

and nucleoli are present. Each nucleoid contains DNA (Fig. 6(a)) and, as we shall see in a later Unit, the enzymes essential for its replication. The number of nucleoids present in a bacterial cell varies according to species and the culture conditions employed. The number of nuclei present in other cell types is usually invariable for that cell type.

ITQ 3 With reference to the cytoplasm (i.e. the area other than the nucleus or nucleoids) in each electron micrograph (Fig. 6), what obvious differences can you see?

Although we cannot use the Feulgen staining technique for electron microscopy, DNA-containing material (*chromatin*) can be resolved under the electron microscope as dark patches within the plant and animal cells (see Fig. 7, which shows a variety of cells, *all* in the process of dividing).

chromatin

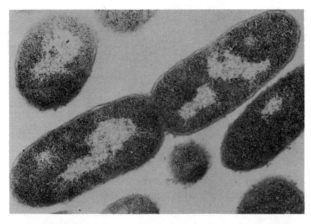

Figure 7(a) Electron micrograph of an *E. coli* cell in the process of dividing. (×23 000)

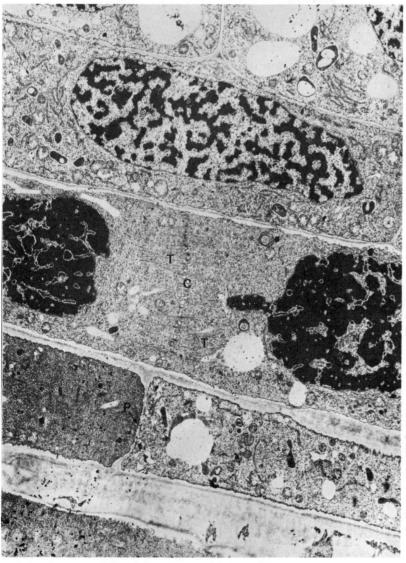

Figure 7(b) Electron micrograph of a group of wheat cells in the process of dividing. T—microtubules; C—cell plate; P—plasmodesmata. (×6 000)

Figure 7(c) Electron micrograph of a cell from a rat ovary in the process of dividing. (× 2 500)

ITQ 4 What changes are apparent in the structure of the nucleoids or the nuclei in these micrographs (Fig. 7) compared with those shown in Figure 6?

QUESTION Why do the chromosomes in these electron micrographs appear more diffuse and incomplete at times, compared with those seen at the same stage by the light microscope? (Hint: chromosomes are coiled three-dimensional structures.)

ANSWER The sections required for the electron microscope are thin and the coiled chromosomes are not cut in every plane of their structure.

What we have said about the structural organization of *E. coli* from Figures 5, 6 and 7 is generally true of all bacteria and blue-green algae. These organisms, which lack chromosomes, are referred to as *prokaryotic* cells. In contrast, the cells of all other plants and animals are referred to as *eukaryotic* cells and the organisms as *eukaryotes*. Viruses do not readily fall into either category as they are not cellular and require a host cell before they can replicate. It is debatable whether they are even living organisms.

prokaryotes
eukaryotes

Table 2 Differences between prokaryotic and eukaryotic cells

Feature	Prokaryote	Eukaryote
nuclear membrane	absent	present
nucleolus	absent	present
DNA	'naked' duplex (double helix)	associated with histone
mitosis and meiosis	absent	present
spindle	absent	present
enzymes of respiratory and photosynthetic electron-transport system	localized on cell membrane	present on membranes of special organelles (e.g. mitochondria and chloroplasts)
ribosomes	70S (30S + 50S)* (ribosomal sub-units smaller than those of eukaryotes)	80S (60S + 40S) (ribosomal sub-units larger than those of prokaryotes)

* S = Svedberg unit or measure of particle mass.

If *E. coli* is gently lysed (broken open) by treatment with a detergent, the DNA material of the nucleoid(s) can be extracted. The material appears to have no basic protein associated with it, but the dimensions (from autoradiographic studies—see Section 2.4) and physical behaviour of this material in sedimentation gradients, indicate that a single, long, circular molecule of DNA is probably present in each nucleoid. There are several reasons for uncertainty, for example, the DNA is frequently in the process of replicating and the molecule can be easily damaged or broken during extraction. Again, we should stress that although this is thought to be true of other prokaryotic cells also, very few prokaryotes (other than *E. coli* and some other bacteria) have been investigated.

In eukaryotic cells, the chromosomes have other components as well as DNA, in particular, basic proteins called histones. A fuller summary of the differences between prokaryotic and eukaryotic cells is given in Table 2. *For the moment, do not worry if you are not familiar with all the terms; they will be covered either later in this Unit or in subsequent Units.*

2.4 Cell division and chromosomes

Cells arise from pre-existing cells by the division of one cell to form two, or by the fusion of two cells (gametes) at fertilization. Consider the first of these methods.

QUESTION Why do cells divide?

ANSWER Cells are restricted in size because of problems associated with the diffusion of gases, the transport of metabolites (food), with communication and the specialization of function. If growth and differentiation of cells are to take place in multicellular organisms, cells must divide and if unicellular organisms are to reproduce, they must divide.

We have already seen that chromosomes are visible in cells only at certain times, yet when they are visible, the somatic, diploid cells of the same individual (and the same species) have the same complement of chromosomes. We are, therefore, led to ask two questions.

1 How is the constancy of the chromosome complement maintained during cell division?

2 Why does staining not reveal chromosomes in *all* cells at all times?

To answer these questions, let us return to a point made earlier when we were discussing methods of locating DNA in cells. Although the Feulgen-staining technique is useful in many respects, it will stain all forms of DNA (i.e. the 'old' DNA and newly synthesized DNA) wherever DNA is present in cells. The alternative technique of using radioactively labelled precursor molecules of DNA should tell us more about *when* DNA is synthesized during the life span of the cell, something about how it is synthesized and the rate at which it is synthesized.

The use of radioactive isotopes of certain common elements (e.g. hydrogen, oxygen, nitrogen, carbon) or radioactively labelled precursor molecules (e.g. molecules in which one of the hydrogen or one of the carbon atoms is labelled), coupled with autoradiographic monitoring techniques, has played a significant role in the elucidation of many metabolic pathways in cells.

Autoradiographic techniques work on a principle similar to that employed in photography. Cells or molecules that have incorporated radioactive precursor molecules are covered with a layer of photographic emulsion and kept in the dark for several days. The emulsion is then developed and fixed like an ordinary photographic negative. Any blackened silver grains on the emulsion arise as a result of radioactive emissions (β-particles from the tritium atoms) and can be located and counted with considerable accuracy.

autoradiography

QUESTION The autoradiographic method, however, does rely on certain basic assumptions—what are they?

ANSWER The method assumes that once the labelled precursor is incorporated into DNA (or RNA), such radioactive atoms will not be lost through the transference of label to other molecules by exchange or catabolism.

ITQ 5 To locate DNA we can use tritiated thymidine (^{3}H-thymidine), and for locating RNA in the cell we can use ^{3}H-uracil. Why are these two nucleotide precursors preferable to ^{3}H-adenine or ^{3}H-cytosine?

2.4.1 Replication and segregation of genetic material in prokaryotes

Let us explore the use of radioactive isotopes and autoradiography to study the way DNA replicates in bacteria. The existence of DNA as a 'naked' double helix, the absence of a nuclear membrane and the complication of mitosis (see Table 2 on p. 64) are all properties of prokaryotes that facilitate easy uptake and incorporation of isotopes into DNA. We shall consider two types of experiment.

1 A recapitulation of the Meselson–Stahl (1958) experiment (S100^2)

In this experiment *E. coli* bacteria were grown in liquid cultures in which the source of nitrogen (as ammonium chloride—NH_4Cl) was the heavy isotope ^{15}N (i.e. 'heavy' compared with the normal isotope ^{14}N). The culture was started ('inoculated') with only a few hundred bacteria, but by the time the numbers had reached many millions (division occurs every 20 minutes or so), the DNA in the bacterial cells contained nitrogen that was almost exclusively ^{15}N. Samples of these bacteria were then transferred to fresh medium for continued growth, but this time the nitrogen source in the culture medium was ^{14}N. Samples of the bacteria were taken immediately, and then after one, two, three and four generations in the new medium, and the DNA was extracted. These DNA extracts were then centrifuged in a caesium chloride (CsCl) gradient and the bands of DNA of different densities were located by noting their absorption of ultraviolet light. These bands and the densitometer tracings, which represent the amount of DNA in each band, appear in Figure 8.

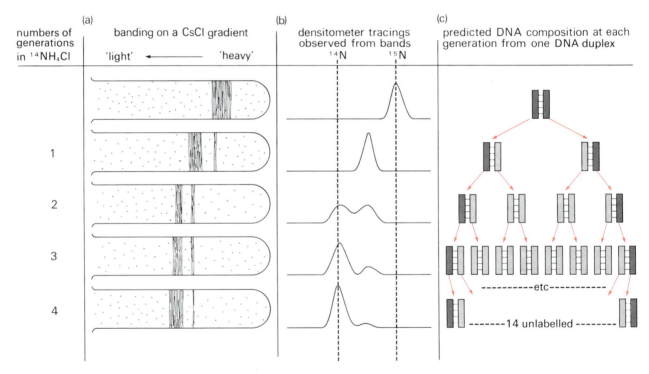

At the beginning of the experiment all the DNA of the bacteria banded in one place, which corresponds to the characteristic position of the heavy isotope in the gradient. After one generation there was again one band, but it was located in a position equivalent to that halfway between the ^{15}N and ^{14}N bands. After two and three generations this band remained, but in decreasing amounts, and a new band appeared, equivalent to ^{14}N DNA; this band became more intense as divisions continued.

The formation of the hybrid $^{15}N/^{14}N$ DNA molecules after one generation and their perpetuation are the key to the mode of DNA replication. The experiments also bring out another important point, namely that one round of DNA replication is completed before the next round is started; there is no overlap between one round and the next. But do these data necessarily represent the way in which whole DNA

Figure 8 E. coli in liquid culture containing $^{15}NH_4Cl$ for several generations and then transferred to $^{14}NH_4Cl$. The dark DNA strands contain ^{15}N; the light strands contain ^{14}N. Hydrogen-bonding between strands is represented by four connecting lines.

molecules replicate? Apparently not, because much fragmentation of the DNA could occur during its extraction from the bacteria. We shall now turn to an experiment that gets over this problem.

2   The Cairns (1961) autoradiography experiment

E. coli bacteria were used again, but this time the replicating DNA from individual bacteria was extracted and observed directly. This is how it was done. The bacteria used required thymidine for their growth. Initially, growth was in liquid culture containing unlabelled thymidine (or ^{2}H-thymidine), then the bacteria were transferred to the same medium but containing labelled thymidine (^{3}H-thymidine) for a time equivalent to nearly two generations. The bacteria were then treated with the enzyme lysozyme, which partially digested their cell walls and as a result gently released their DNA. Autoradiographs of this DNA were then made (Fig. 9).

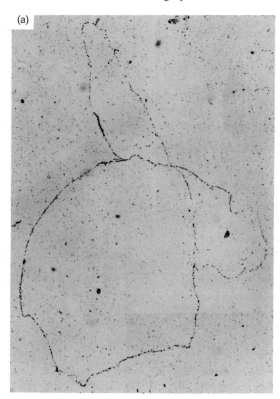

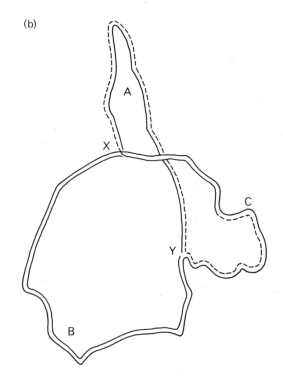

QUESTION How do you account for the fact that in Figure 9 the section C to Y and X to A to Y (dashed lines in Fig. 9(b)) appear different from the remainder of the DNA which is autoradiographed?

ANSWER The C to Y section contains a strand of labelled and a strand of unlabelled DNA (i.e. ^{3}H/^{2}H regions), which, therefore, gives rise to lines of blackened grains at half the density of the X to A to Y section where both strands of DNA are labelled (i.e. ^{3}H/^{3}H region).

Figure 9 (a) An autoradiograph of replicating DNA from *E. coli*; (b) a diagrammatic interpretation. A bold line signifies a single strand of DNA containing labelled ^{3}H-thymidine. The broken line represents the position of an unlabelled single strand of DNA, which will not be visible on the autoradiograph.

The point is further demonstrated by the grain counts made by Cairns (Table 3) for the three sections (A, B and C) shown in Figure 9.

Table 3   Grain counts from three sections from the autoradiograph in Figure 9

Section	Number of grains	Length of section (μm)	Grains (μm)
A	714	670	1.1
B	1 298	680	1.9
Y to C	213	215	1.0
C			
C to X	359	205	1.8

As you may find this autoradiograph (Fig. 9) rather difficult to understand, let us just spend a little more time in its explanation.

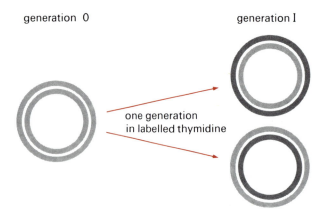

one generation
in labelled thymidine

Figure 10

After the growth of *E. coli* in the unlabelled thymidine we predict the behaviour of the two strands, which can be represented diagrammatically by two light grey circles (see Fig. 10, generation 0).

After one generation in labelled thymidine, two new bacterial cells are produced, each containing one labelled (dark grey) and one unlabelled (light grey) strand of DNA, as in Figure 10, generation I. What the autoradiograph (Fig. 9) shows is one one of these DNA molecules nearing the end of the second generation (replication) in labelled thymidine, so that loop Y–A–X has one strand labelled and one un-labelled, whereas loop X–B–Y has both strands labelled. The double labelling in section C to X can be explained by the fact that at the time of changing the bacterium from the unlabelled to the labelled medium, the DNA was just completing DNA replication and, therefore, picked up the label in this region in one strand, just before the first full round of replication.

Thus, it can be deduced that hybrid DNA (i.e. labelled/unlabelled) *is* formed during the replication of intact DNA molecules. Furthermore, the DNA molecule of *E. coli* is in the form of a circle, and what you see in Figure 9 is a replicating circle with replicating forks at X and Y formed as a result of replication that starts at a single fixed point. Originally, it was thought that replication proceeded in only one direction; there is now good evidence that replication may proceed from one point in two directions. This is true not only for the DNA of *E. coli* but also for other bacteria.

> **ITQ 6** The type of DNA replication consistent with these results is termed 'semi-conservative'. Can you explain this term in the light of the two experiments we have just outlined?

Evidence for the *semi-conservative replication* of DNA in bacterial cells provided support for the Watson and Crick hypothesis put forward earlier, in 1953 (see *HIST*). They had assumed that the uncoiling and separation of the two chains would occur progressively from one end of the molecule (like the opening of a zip-fastener) and that new synthesis would begin to occur straight away, yet progressively, along the molecule from the same physical end in both chains. Despite the apparent simplicity, the synthesis of new DNA chains still presents us today with a very challenging problem. The problem arises because the DNA strands run in an opposite direction (Fig. 11). Reading downwards, the sequence of atoms in one strand is . . . 3′C, 4′C, 5′C, O, P, O, 3′C, 4′C, 5′C . . . (3′ to 5′ direction) and in the other, the converse (5′ to 3′ direction), where 3′, 4′ and 5′ refer to specific carbon atoms in the deoxyribose molecule.

semi-conservative replication

The DNA polymerases that have been found (i.e. the enzymes involved in DNA synthesis) all have at least one property in common: they cause chain growth *only* in the 5′ to 3′ direction. No enzyme has been isolated that is capable of bringing about chain growth in the 3′ to 5′ direction. Are both chains, therefore, being used in replication, and if so, how?

We mention this problem for two reasons. First, despite all the evidence supporting the semi-conservative replication of DNA, we are still not completely clear how this is accomplished in the cell, although there is evidence that the 3′ to 5′ chain grows via the formation of small fragments synthesized in the 5′ to 3′ direction and subsequently joined together. Secondly, the technical problems become even more involved when we move from the simple 'chromosome' of prokaryotes to the

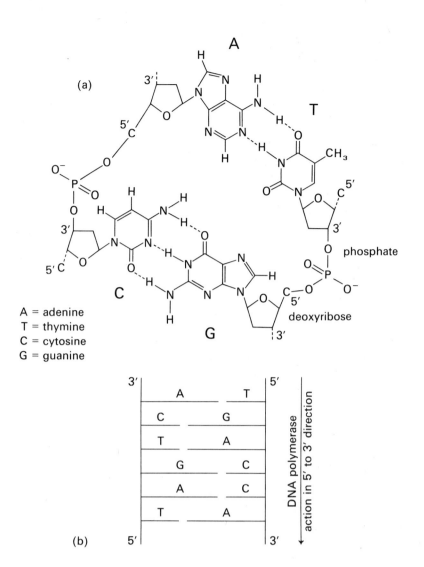

A = adenine
T = thymine
C = cytosine
G = guanine

Figure 11 (a) Antiparallel strands of DNA; (b) the direction of DNA polymerase action. 3′ and 5′ refer to specific carbon atoms in the sugar deoxyribose.

chromosomes of eukaryotes. Nevertheless, it seems likely that chromosomal DNA synthesis is also semi-conservative. Evidence comes from the experiments of Taylor, Woods and Hughes in 1957 and others subsequently. Here again, the use of radio-active labels has proved most valuable. Taylor and his colleagues incubated the rapidly dividing cells in the root tip of the broad-bean (*Vicia faba*) with ^{3}H-thymidine. Their choice of the broad-bean was convenient not only for its ready availability, but also because it is an organism with large but relatively few chromosomes (12).

QUESTION Why are these two features advantageous in chromosomal studies?

ANSWER The selection of organisms whose cells contain large but relatively few chromosomes enables more accurate observations about chromosome behaviour to be made because often each chromosome can be readily distinguished and quantitatively measured with respect to any other chromosome present. In fact, ^{3}H-thymidine studies are limited to organisms with large chromosomes because the β-particles from the ^{3}H penetrate about 1 μm into photographic emulsion, and hence structures smaller than this cannot be resolved.

Taylor and his colleagues also chose these cells because previously other workers had established the time of DNA replication in relation to mitosis. The cells were allowed to go through either one or two full generations, before they were 'fixed' at metaphase. To remind you of the essentials, a summary of the main phases of the mitotic cycle are given in Figure 12 (see also S100[3]).

Mitosis is a mechanical process that ensures that sister chromatids separate accurately and in a co-ordinated way. The duration of mitosis varies between species, and indeed within species, under different external conditions; but, as we shall see in Section 2.6, it is only one relatively short phase in the total cell cycle. Those cells that are not undergoing mitosis are said to be in *interphase*.

chromatids

interphase

69

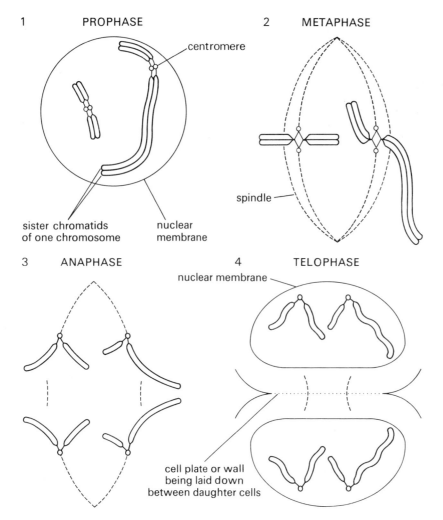

1 PROPHASE

centromere

sister chromatids
of one chromosome

nuclear
membrane

2 METAPHASE

spindle

3 ANAPHASE

4 TELOPHASE

nuclear membrane

cell plate or wall
being laid down
between daughter cells

Figure 12 The main stages of mitosis.

During the early stages of mitosis (*prophase*), the chromosomes become visible as discrete, compact bodies, because of their internal coiling and condensing. The compact structure, which is thought to involve configurational changes in the DNA molecule as well as a DNA–protein interaction, facilitates the organization and separation of the chromosomes during the mitotic phase. During prophase, there are also other significant events. The nuclear membrane breaks down, the nucleolus becomes indistinct and a system of microtubules is in evidence, as shown (arrowed) in the electron micrograph (Fig. 13, *opposite*). These microtubules eventually constitute the spindle.

The microtubular proteins of the spindle appear to be synthesized during interphase, although the organization of the spindle itself occurs after the nuclear membrane has broken down, when the chromosomes have almost reached their greatest degree of compactness. The spindle is difficult to see, except with a polarizing light microscope or an electron microscope, but with these microscopes it is possible to detect that the spindle consists of continuous fibres that extend from pole to pole. Soon the compact, densely staining, chromosomes attach themselves to the equator of the spindle by a special constricted region known as the centromere. This phase in mitosis is referred to as *metaphase*. At this point in division the sister chromatids are quite separate, except for their attachment at the centromere. A little while later, the daughter chromatids separate and move quite quickly to their respective poles of the spindle (*anaphase*). The two groups of single stranded chromosomes uncoil (*telophase*) and, with the re-formation of nuclear membranes and the laying down of a cell wall (plants) or pinching off of the two halves of the cytoplasm (animal), revert to the interphase state. Mitotic cell division is now complete.

ITQ 7 From this description and Figure 12, identify the various stages of mitosis in the diagrams that follow (Fig. 14, *opposite*). (Note: the pictures are not necessarily in the correct chronological sequence, but they are of the same magnification.)

The diagrams are labelled A–H. Match each letter with a number so that 1 represents the earliest stage in mitosis and 8 the latest.

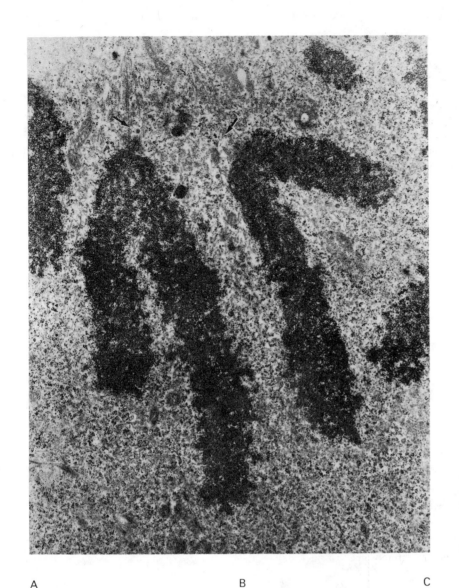

Figure 13 Electron micrograph showing two wallaby chromosomes in early anaphase. Note particularly the micro-tubule attachments to the centromeric region of each chromosome (arrowed). We can distinguish individual micro-tubules in this electron micrograph, but they cannot be seen in the light micro-scope. The spindle fibres that can be resolved by the light micrograph are in fact bundles of microtubules. ($\times 15\,000$)

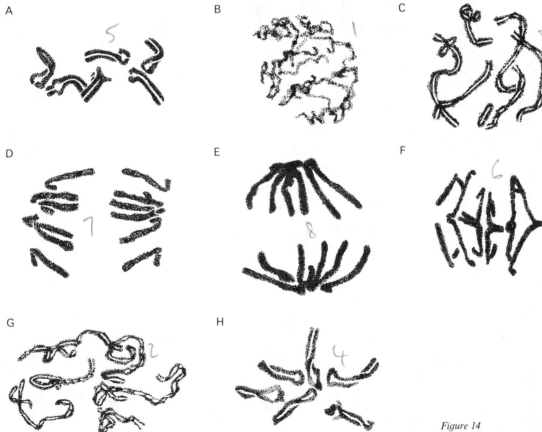

Figure 14

With the information about mitosis fresh in your minds, let us return to Taylor's experiments.

In addition to the radioactive precursor, Taylor and his colleagues included the drug colchicine in their incubation medium. Colchicine interferes with the formation of spindles without preventing chromosome replication with the result that the number of chromosomes is doubled at each subsequent division. Colchicine also has the advantage of causing the chromatids in each chromosome to diverge, so that they are held together only at the centromeric region. The colchicine-treated cells obtained by Taylor were then fixed, stained (Feulgen) and autoradiographed. The results are shown in Figures 15 and 16. (For clarity only 5 of the 12 chromosomes are shown in these Figures.)

Figure 15 (*left*) Autoradiographed chromosomes after one replication cycle in which ³H-thymidine was incorporated into DNA.

Figure 16 (*right*) Autoradiographed chromosomes after one replication cycle with ³H-thymidine, followed by replication again in the absence of labelled thymidine (i.e. with only unlabelled thymidine present). Note that in one or two sister chromatids (arrowed) there appears to have been some exchange of DNA, which we shall discuss later in this Section.

QUESTION Assuming that semi-conservative replication does take place, how can the results obtained by Taylor *et al.* (Figs. 15 and 16) be explained diagrammatically?

ANSWER Your diagram should resemble Figure 17. We have represented the unlabelled strands by broken lines and the labelled strands by continuous lines.

If the DNA is contained within the chromosome as a single duplex, then:

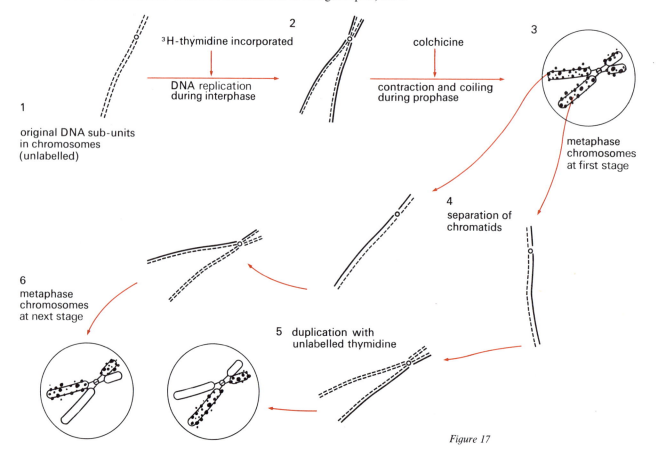

Figure 17

72

ITQ 8 In a few of the metaphases after 34 hours in colchicine (one cycle in ³H-thymidine and two cycles in unlabelled thymidine), Taylor and his colleagues found 48 chromosomes in each cell, indicating two successive mitoses (12 → 24 → 48). How would you expect the label to be distributed among the chromatids if semi-conservative replication of DNA in the chromosomes had taken place?

At this stage, note that although Taylor's experiments show that chromosomes replicate semi-conservatively and are compatible with the hypothesis that chromosomal DNA replicates semi-conservatively, they do not prove it, as the experiments are open to other interpretations.

ITQ 9 Which of the following statements are compatible with the results of Taylor *et al.*?

(a) Each chromosome contains a single giant DNA molecule that replicates semi-conservatively.

(b) The chromosome is composed of two sub-units, one old and one new, which extend the whole length of the chromosome, so that the DNA of each sub-unit is conserved.

(c) The chromosome contains a number of DNA molecules, each replicating semi-conservatively and held together in such a way that the new nucleotide chains form a unit, which at the next replication separates from the unit formed by the old chains.

(d) The chromosome sub-units contain one or a number of DNA molecules that replicate conservatively, that is, maintain the whole molecular structure through the replication process; the new molecule or molecules form a unit that at the next replication separates from the unit formed by the old molecule or molecules.

One other feature that Taylor and his colleagues observed (mentioned in the answer to the last QUESTION and ANSWER and arrowed in Fig. 16) was that, in some metaphase chromosomes, after two replication cycles there was an exchange of segments between the two chromatids. Sometimes the exchanges were single, sometimes double (i.e. twin exchanges at corresponding positions in sister chromosomes and therefore probably reflecting a single exchange at the earlier division). From the relative frequency of single and twin exchanges, they concluded that the DNA sub-units of the chromosomes are structurally different and that reunion can occur only between those with similar structures. Taken in conjunction with the semi-conservative replication, which the chromosomes apparently showed, it appeared highly probable that the DNA in the chromosome was also replicating semi-conservatively. This view has since been much strengthened by the evidence, from yeast and *Drosophila*, that DNA molecules as large as whole chromosomes can be extracted from nuclei; hence, the chromosome is essentially a single giant DNA molecule.

In contrast to DNA, histones are present in more variable amounts in different cells of the same species; also, according to studies with radioactively labelled amino acids, chromatid proteins and RNA are not conserved from generation to generation.

2.5 The cell cycle

Having established chromosome replication in terms of DNA synthesis, we can now turn to another point, namely the constant relationship between the amount of DNA and the number of sets of chromosomes in the nucleus. In dividing cells, the quantity of DNA per nucleus doubles during interphase—in the so-called S (synthesis) phase*—and is then *precisely* and *equally* distributed between the two groups of chromosomes at anaphase or telophase, before cell division.

---

* NB In describing the cell cycle, the terms 'phase' and 'period' tend to be used interchangibly.

The cell cycle showing the sequential relationship between DNA synthesis and mitosis is given in Figure 18. The phases in the total cycle are represented by very approximate arcs on the circle.

cell cycle
phases—S, G_1, G_2

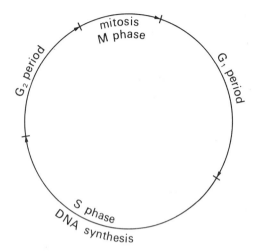

	% total cell cycle time
M = mitosis	5-12
G_1 = gap between end of mitosis and onset of DNA synthesis	20-40
S = gap between end of DNA synthesis and mitosis	20-50
G_2 = gap between end of DNA synthesis and mitosis	10-30

interphase { G_1, S, G_2 }

Figure 18 Phases of the cell cycle.

You will notice that in this generalized scheme, the percentages given are very variable. The variation occurs because of differences among species, and differences even within species when environmental conditions such as temperature and food supplies change.

How then have the relative durations of these phases in the cell cycle been determined?

QUESTION Given some actively dividing group of cells, for example, broad-bean root tips, and a suitable radioactively labelled precursor solution, how would you determine the period of DNA synthesis in the cell cycle?

ANSWER By immersing the roots in the radioactive solution at time zero and then removing equal-sized samples of root tips at intervals to estimate when the isotopic label is incorporated into the chromosomal DNA.

The earliest work of this kind was carried out in the 1950s using broad-bean root tips and a radioactive solution containing ^{32}P. Today, ^{3}H-thymidine is preferred because ^{32}P is not a very specific precursor of DNA. Consequently, it will be incorporated not only into DNA, but also into many other molecules within the cell (e.g. ATP, phosphates, etc.); this could create ambiguities in the interpretation of autoradiographs.

pulse-labelling

The method principally used nowadays is called *pulse-labelling*. It consists of incubating the cells for a short period (about 30 minutes) with ^{3}H-thymidine and then removing the radioactive label from the incubating medium. After this, the cells are grown in unlabelled medium and then sampled at intervals thereafter; the proportion of mitoses (scored at metaphase), which are labelled in autoradiographs, are counted. A typical result is shown in Figure 19 (*opposite*).

Examine Figures 18 and 19 carefully, then work through the following QUESTION *and* ANSWER. *Remember that not all the cells divide simultaneously (i.e. they are not synchronized).*

QUESTION Referring to Figure 18 explain what is happening during the intervals labelled 1 to 4 inclusive on the graph in Figure 19.

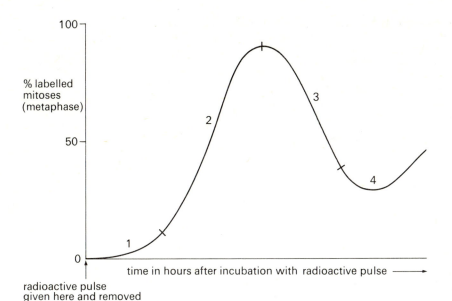

Figure 19 The number of dividing cells, scored at metaphase, showing the presence of a radioactive label. Readings were taken at various times after the pulse label was given.

ANSWER *Interval 1* Assuming that there is only a short delay before the labelled ³H-thymidine is taken up, the earliest samples will have very few labelled mitoses because those cells already starting mitosis are not synthesizing DNA (they have already synthesized it!).

Interval 2 The proportion of labelled mitoses rises steeply to a maximum as the cells that were in the S phase at the time of giving the labelled pulse come through to division.

Interval 3 Following the peak there is a fairly sharp drop in the number of labelled mitoses as the cells originally in the G_1 period come to the end of their cycle.

Interval 4 represents the remainder of the cells originally in the G_1 period or in the anaphase or telophase of mitosis.

The average S period is taken as the time between the two points in the first wave where 50 per cent of the mitoses are labelled (Fig. 20).

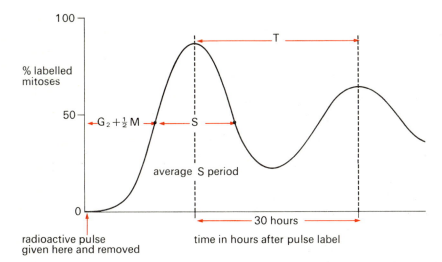

Figure 20 The total cell-cycle time is here superimposed on Figure 19 and the principal stages in it are indicated (see Fig. 18).

The $G_2 + \frac{1}{2}M$ phase is also shown in Figure 20. Half the mitotic time is added to G_2 because the mitoses are scored in metaphase, and prophase takes approximately half the time taken for mitosis. The total time taken to complete a cycle T (i.e. between two peaks) enables us to estimate $G_1 + \frac{1}{2}M$, because $G_1 + \frac{1}{2}M = T - S - G_2 - \frac{1}{2}M$.

Also, the time taken for *mitosis alone* can be quite accurately determined by time-lapse cine photomicrography.

An approximate time scale for the complete cell cycle in root-tip cells of *Vicia faba* at 18 °C is given below.

$$\begin{array}{ll}
\text{total cell-cycle time} & T = 30 \text{ hours} \\
\end{array}$$

$$\left.\begin{array}{ll}
G_1 & 12 \text{ hours} \\
S & 6\text{--}7 \text{ hours} \\
G_2 & 8 \text{ hours} \\
\end{array}\right\} \text{ interphase } 26\text{--}27 \text{ hours}$$

$$\left.\begin{array}{ll}
\text{prophase} & 1\frac{1}{2} \text{ hours} \\
\text{metaphase} & 1\frac{1}{2} \text{ hours} \\
\text{anaphase} & \frac{1}{4} \text{ hour} \\
\text{telophase} & \frac{3}{4} \text{ hour} \\
\end{array}\right\} \text{ mitosis 4 hours}$$

These values are only approximate for this species at this temperature. Other species show longer or shorter total times for completing the cell cycle and variable G_1, S, G_2 and M times.

2.6 The salient features of mitosis and the cell cycle

At this point we should summarize the main features of somatic cell division in eukaryotes.

1 It is important to view the process of somatic cell division (*mitosis*), as a dynamic process; it is divided into prophase, metaphase, anaphase and telophase for descriptive convenience. Essentially, mitosis ensures the accurate yet co-ordinated separation of sister chromatids (and hence DNA with its coded information). Mitosis, therefore, maintains a constant complement of chromosomes, provided that there is accurate replication of the DNA and accurate duplication and separation of the chromosomes between one generation and the next.

2 Mitosis must be considered as an integral part of the complete cell cycle. Mitosis is the mechanical separation of the nuclear contents, a stage where the chromosomes are visible; whereas *interphase* is a mechanically inactive stage, in which the chromosomes are not visible. Yet it is during this metabolically active stage (the S phase) that DNA synthesis takes place and the original amount of DNA is doubled. As a result, when the chromosomes first become visible at the prophase of mitosis, they are double along their entire length, even at the centromere. After mitosis, however, the amount of DNA is divided equally between the two daughter cells.

2.7 Meiosis

Chromosomes are found in all eukaryotic cells and chemical analyses of the DNA in different somatic cells of the same organism reveal a remarkable constancy of amount; through the mitotic phase of the cell cycle, the chromosomal content is normally replicated accurately and the chromosomal complement is kept constant.

However, there is one extremely important phenomenon which we still have not explained. In organisms that reproduce sexually, a new organism is produced as a result of the fusion of two cells (egg and sperm cells). Yet, observations show that there is normally a remarkable constancy in the chromosome number in individuals of the same species from generation to generation. The observed constancy can be explained only if there is at some point a reduction in the chromosome complement. But when does this occur?

We can get an answer to this question by studying Table 4, which sets out the chemically estimated amounts of DNA in sperm cells and red blood cells for a variety of vertebrates.

QUESTION What general inference can be drawn from Table 4?

ANSWER The sperm cells contain half the amount of DNA (within experimental accuracy) found in the red blood cells.

QUESTION Would you expect the egg cells also to contain the same amount of DNA as the sperm cells? If so, why?

ANSWER Yes. This is an argument about symmetry—by this means the DNA (and hence by deduction the chromosomal) complement would remain constant from generation to generation.

Table 4   The mean values for DNA content in sperm cells and red blood cells of various organisms, in picograms (10^{-12} g)

Organism	Sperm cell	Red blood cell
carp	1.64	3.49
brown trout	2.67	5.79
toad	3.70	7.33
chicken	1.26	2.49
man	3.25	7.30*

* The red blood cells in man were taken from the bone marrow because the mature red blood cells circulating in the blood (erythrocytes) do not possess a nucleus.

Although we cannot deduce anything about the amount of DNA per cell in relation to the size and complexity of individual organisms from this Table, all the gametes (sperm and egg) have half the DNA content of the somatic cells. We might tentatively predict from this that the chromosome complement in gametic cells is *haploid*, whereas in somatic cells it is *diploid*.

But how does the reduction from the diploid to the haploid state come about?

A clue to the problem was found very early in the history of genetics, in the 1880s by van Beneden. Working with the horse roundworm *Parascaris equorum*, he noticed that the egg and sperm nuclei each contained two chromosomes, whereas at mitosis in the fertilized egg, four chromosomes were clearly visible. (Note the great advantage of working with an organism that has few chromosomes!) Later, Weismann (1887) (see *HIST*) clearly predicted that there must be a special nuclear division, repeated in every generation of the organism, whereby the chromosome number is reduced to half the number contained in the parent nucleus. If Weismann's predictions were true, then there were likely to be two sets of *homologous chromosomes* (a set from each parent) in every diploid cell, but only a single set of chromosomes in a haploid cell.

homologous chromosomes

By the end of the nineteenth century, Weismann's predictions were given substance, as details of meiotic cell division or *meiosis* were worked out. Essentially, meiosis is a process in which two nuclear divisions take place, but during this time the chromosomes divide only once; hence it is often called *reduction division*.

Figure 21   Some of the main stages in meiosis.

The main stages of meiosis are shown in Figure 21 and summarized overleaf.

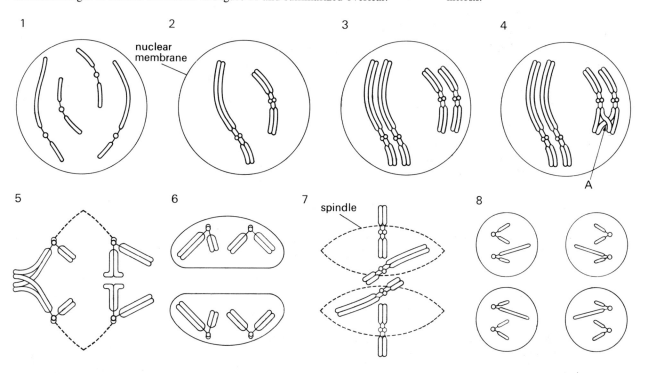

1   The diploid number of chromosomes appear as single threads (compare this with mitosis, where they are double).

2   Homologous chromosomes (i.e. the two sets derived from the two parents) pair specifically side-by-side, centromere to centromere and, as we shall see in Unit 3 and TV programme 3, allele to allele. Each pair is called a *bivalent* and the process is spoken of as pairing or *synapsis*. Such pairing is possible because, as the drawing from an electron micrograph (Fig. 22(b)) shows, a synaptinemal complex of protein filaments forms between each synapsed pair of homologous chromosomes.

**bivalent
synapsis**

(a)

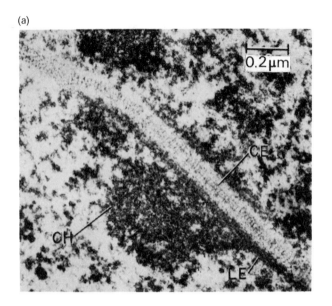

(b)

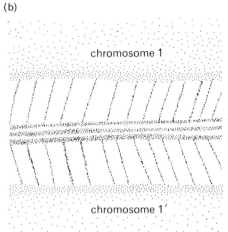

Figure 22 The synaptinemal complex lying between the two synapsed chromosomes. (a) An electron micrograph. CE—central element; CH—chromatin of chromosome 1′; LE—lateral element. (b) A drawing from an electron micrograph.

3   Each chromosome divides into two sister chromatids. Each chromatid keeps in close proximity to its sister.

4   'Breaks' (induced by enzymes) occur here and there at intervals along the chromatids, but any particular 'break' affects only one of the sister chromatids at any one point. A 'break', however, occurs at corresponding positions in the homologues.

Broken ends join up cross-wise to form *chiasmata* in new combinations (see arrow A in Fig. 21). Thus, an exchange of genetic material occurs; its significance is discussed in Unit 3.

chiasmata

5   The nuclear membrane breaks down and a spindle is formed. The bivalents move to the equator of the spindle in such a way that homologous centromeres come to lie equidistant from the equator and equidistant from their respective poles. The homologous centromeres then move unseparated to opposite poles of the spindle.

6   The spindle disappears and nuclear membranes may or may not form around each daughter group. This ends the first or *reduction division* stage of meiosis, where the chromosome number is effectively reduced by a half. There is then usually an interphase period.

7   The second division stage of meiosis begins with the formation of new spindles in each daughter cell (the spindles are usually at right angles to the *reduction division* stage spindle).

The centromeres move to the equators of each new spindle, followed shortly after by the sister half-centromeres (and chromatids) moving to opposite poles of the spindle.

8   A nuclear membrane is organized around each of the four groups of chromatids.

   ITQ 10 Figure 23 shows various stages (labelled A–H) in meiosis. The pictures are *not* necessarily in the correct sequence. Match the diagrams with the descriptions (1–8) you have just read.

How did you get on? *If you did not get them right the first time, go back and see where you made your mistake.*

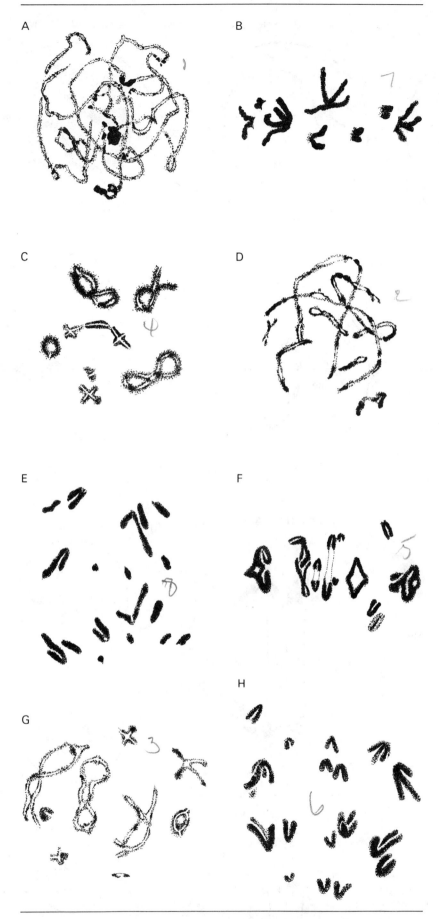

Figure 23 Chromosomes from grass-hopper testis (the magnification is the same in all the photographs).

Cytogeneticists have given specific names to the various stages in the meiotic sequence, as a precise shorthand reference. *You are not expected to learn these names to achieve the objectives of this Unit*, but we shall use them occasionally in this and other Units. For this reason, the various stages are shown in Figure 24 (*overleaf*), which should be used as a source of reference.

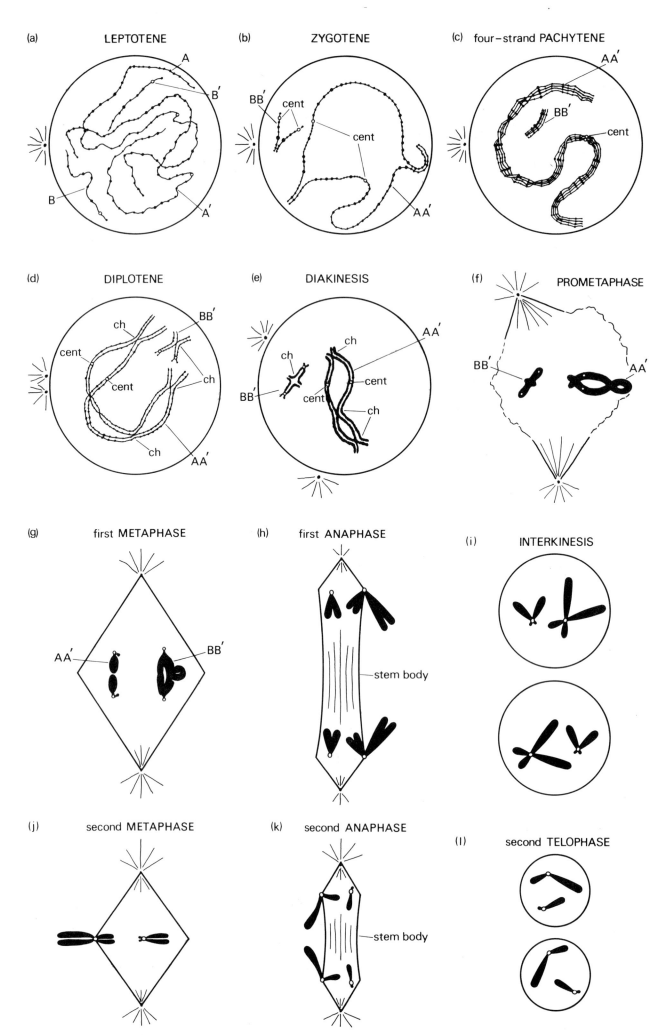

Figure 24 A summary of the main stages of meiosis. A and A′ and B and B′ are pairs of homologous chromosomes. cent—centromere; ch—chiasma.

2.7.1 DNA synthesis and chromosomes

To conclude this Section, we can now relate our findings in mitosis and meiosis to changes in amount of DNA in the nucleus. The amount of DNA present in the haploid cell is equivalent to an amount that we shall call 1C; the amount in a diploid cell is, therefore, 2C. At the end of the S phase in the diploid cell cycle, the 2C amount of DNA is doubled (4C), but this is equally divided at the end of mitosis between each new daughter cell, so that each contains 2C.

The time and pattern of DNA synthesis related to the meiotic cycle is rather different from that related to the mitotic cycle. Although an S phase occurs in the interphase period of the cycle, in which the 2C amount of DNA is doubled to 4C, there is little or no G_2 phase (see Fig. 18). Consequently, DNA synthesis may be completed only at the early leptotene and this may possibly account for the visible split into two chromatids after meiosis has started (compare this with mitosis where the chromatids are quite distinct at early prophase).

In the first meiotic division, the 4C amount of DNA is reduced to 2C; the 2C amount is halved to 1C by the end of the second meiotic division as a result of the distribution of the four chromatids of each kind present initially (two in each homologue; compare this with stage 3 in Fig. 21), one to each of the four nuclei resulting from meiosis. Another marked difference between meiosis and mitosis is that a small amount of DNA synthesis actually occurs during prophase I of meiosis. Some (but not all) of this synthesis is presumed to be concerned with DNA repair at chiasma formation (see Units 3 and 4).

2.8 Recognition of genes

So far in this Unit you have considered the chemical nature of the genetic material and how its constancy is maintained from cell to cell and from generation to generation. At the molecular level, this constancy is maintained by the semi-conservative replication of DNA in prokaryotes and probably in eukaryotes, where semi-conservative replication is thought to occur in the chromosomes. In organisms with a sexual phase that involves fusion of genetic material from two 'parents' at some stage in their life cycle, constancy is ensured by a reduction division, or halving of chromosome numbers, at some stage in that cycle. How do these physical realities relate to the genetic characteristics of organisms, or, putting it another way, how do we recognize genes? To answer these questions we shall turn our attention to the genetic cross. We shall consider organisms other than the garden pea. Also, we shall need to look beyond crosses involving single pairs of alleles to those involving two pairs. We shall run into problems of nomenclature—the individuality of geneticists when it comes to naming genes is only too apparent when you look at the labels they have invented! Finally, the genetic basis of the differences between sexes (essential for making crosses!) will also be discussed.

2.8.1 The genetic cross

First, refresh your memory of Mendel's experiments with the garden pea (Unit 1). You will notice that he first observed differences between his plants and then through breeding experiments established his 'laws' of inheritance. He did so by recognizing alternative forms of the units of genetic information (now called alleles) such as those concerned with the colour of seeds (green or yellow) or the shape of seeds (wrinkled or smooth). The important point is that the units of information (actually called 'factors' by Mendel but later called 'genes') became apparent only because different alleles existed, and the proof of their existence required making crosses between plants with *different* characteristics. The emphasis then is on the necessity of making crosses in order to recognize genes.

In genetics, a cross can be a mating, an infection, a fusion, a pollination or a conjugation (Table 5, *overleaf*), depending on the organism. A geneticist, suspecting that two different variants may each be heritable, attempts to confront one variant with the other or, more accurately, confronts the whole or part of the heritable system of one variant with that of the other. The experiment is set up by the geneticist, but the operation is obligingly done by the organism itself! The geneticist then waits to look at the result in the next and subsequent generations to see whether the original phenotypes can still be recognized and in what numbers they appear. To use the correct terminology, the geneticist is interested to know how they *segregate* with or from each other.

In Table 5 we give you a broad view of the way crosses take place in a wide range of different organisms. You have no need to remember the details at this stage because they will be reiterated and some will be expanded upon in later Units, particularly in Unit 3.

Table 5   The genetic cross

Organism	Fusion event	Segregation event
'higher' animals	fusion of egg and sperm cells to form zygote cell	production of eggs and sperm
'higher' plants	fusion of nucleus in pollen tube with nucleus in ovule to form zygote (seed)	production of ovules and pollen
fungi	fusion of hyphae, spores, or spores and hyphae, and then fusion of nuclei	formation of sexual spores
bacteria	conjugation transduction transformation } fusion of part of one genome with another whole genome	production of each bacterial clone (after conjugation or transduction or transformation)
viruses	mixed infection	production of virus particles from the 'burst'

The next step is the crucial one, and this was made successfully for the first time by Mendel. It is to rationalize the numbers and kinds of phenotypes that are segregating by suggesting hypothetical segregating factors (genes), responsible for the difference in phenotype.

2.8.2   The gene defined by segregation analysis

Fungal spores from wild-type *Neurospora crassa*, a haploid organism, germinate and grow on a minimal* medium and are termed prototrophic. By contrast, spores from certain strains (*auxotrophs*) grow only when the medium is supplemented in some way. The colonies of these strains maintain their auxotrophy through prolonged growth and sub-culture, that is, they possess heritable differences. Suppose that a cross is made in which one parent is auxotrophic† by requiring adenine (*ad*) and the other *prototrophic* or wild type (*ad*$^+$). Individual nuclei from each parent fuse to form a diploid zygote nucleus, which then goes through one meiotic and one mitotic round of division to give eight haploid nuclei, each of which forms a sexual spore. Groups of eight spores from such an event can be isolated and dissected one by one and, finally, germinated separately in petri dishes on agar medium containing adenine (Fig. 25).

prototroph
auxotroph

When each spore has germinated to form a colony, a transfer is made from each colony to minimal medium to see whether the colonies grow without adenine. From each and every group of eight spores one finds four spores that need adenine for growth and four that do not. The phenotypes of these colonies are as stable as those of the parental colonies.

So, after the fusion process and the three divisions, we recover what we put in (auxotrophs and prototrophs), and we find equal numbers of each type—again what we put in. The phenotypic difference (growth, or no growth without adenine) has shown *segregation* after fusion. It appears in some spores and not others. We take advantage of such a situation to *define* a gene that is concerned with adenine metabo-

segregation

* A minimal medium is the simplest chemically defined medium on which the wild type of a fungal (or bacterial) species will grow.

† This term should not be confused with autotrophic (see p. 102).

lism. This gene can exist in two forms; the presence of one or the other form determines whether the organism requires adenine for growth or not. The zygote cell must have carried a copy of the genes from both parents to be able to transmit them to the products of division, yet we never find a sexual spore that grows into a colony that is a mixture of hyphae, some growing without adenine and others requiring it. Apparently, the sexual spores each receive only *one or other* copy. As you will recall from Unit 1, we call the alternative forms of the gene *alleles*.

allele

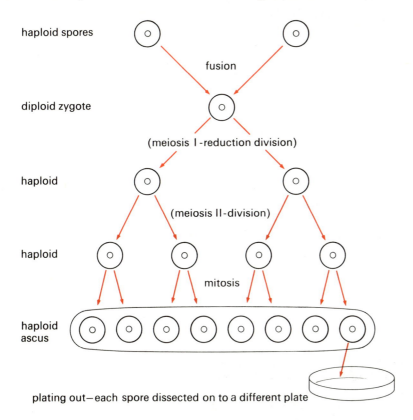

haploid spores

fusion

diploid zygote

(meiosis I -reduction division)

haploid

(meiosis II-division)

haploid

mitosis

haploid
ascus

plating out—each spore dissected on to a different plate

Figure 25 Plating out the meiotic products (haploid spores) from an individual ascus of *Neurospora*.

Consider another cross made with *N. crassa* between a methionine-requiring strain (*met*) and the one with the adenine requirement (*ad*). Sexual spores from the cross are germinated on a medium containing adenine and methionine. Why do we do this? It is to avoid selecting against particular genotypes that might be auxotrophic and would otherwise die. Each colony results from an isolated spore and is transferred to culture medium to see whether it is able to grow without methionine or adenine or without both.

> **ITQ 11** Using the information given, write down the types of colony, defined by their growth requirements, that you would expect this cross to produce, together with the relative proportions of each type you would expect in a sample.

Hopefully, you obtained the correct answer to ITQ 11, that is, that there would be four types formed in equal numbers. But you might have suggested, in answer to the ITQ, that there would be only two types of colony, each resembling one or other of the parents (*ad met*$^+$ and *ad*$^+$ *met*). Do not worry if you did not predict the other two classes, but you should remember it on a future occasion.

The following Table shows actual data from such an experiment.

Table 6   Data from a cross between *ad met*$^+$ and *ad*$^+$ *met*

Colony phenotype	Colony genotype	Number of colonies
requires adenine and methionine	*ad met*	27
requires nothing	*ad*$^+$ *met*$^+$	30
requires adenine	*ad met*$^+$	36
requires methionine	*ad*$^+$ *met*	38

What can we deduce from these data?

First, we can deduce that both the gene for the methionine requirement and the gene for the adenine requirement are segregating, so we conclude that a second gene is manifesting itself. We could also put this conclusion another way, and say that the results show that *met* and *ad* are not manifestations of the same gene, that is, they are non-allelic. By definition, however, *met*$^+$ and *met* are alleles because the meiotic (haploid) product will have either *met*$^+$ or *met*, but *never both or neither*. Consequently, *met*$^+$ and *ad*$^+$ are *not* allelic because, in addition to the apparent segregation products (*met*$^+$ *ad* and *met ad*$^+$), there are meiotic products with both (*met*$^+$ *ad*$^+$) and meiotic products with neither (*met ad*), and by definition these cannot be allelic. We also see from these experiments the great advantage of working with haploid ascospores, because each ascospore gives rise to a haploid organism whose phenotype directly reflects the genotype of the ascospore.

Secondly, we see that the form of the *met* gene (the allele—*met*$^+$ or *met*) appears with *ad*$^+$ and *ad* in about equal proportions. We say that the two genes are assorting independently from each other or that *independent assortment* is occurring.

independent assortment

The numbers of the four types of colony were not *exactly* equal. We should not shrug off discrepancies as 'near enough'; instead, we should use a test that will tell us how often we should expect random fluctuations as large as those seen from an exact $1:1:1:1$ ratio for that size of sample. This test is the χ^2 test.

If you are in doubt how to apply the χ^2 test, consult STATS, Section ST.4.

ITQ 12 Work out the χ^2 value for the expectation that all four classes are of equal frequency (i.e. 131/4).

2.8.3 Genetic symbols and notation

We have used the symbols *ad* and *ad*$^+$ and *met* and *met*$^+$ to denote pairs of alleles. The use of $+$ and $-$ signs to denote either the presence or absence of the ability to synthesize adenine or methionine might seem more consistent, but we have opted for the most common usage. Unfortunately, there is no standard notation shared by geneticists who work with different organisms. One of the reasons for this diversity is that, although there is general agreement that a plus sign should be used for the genotype when it occurs in the normal or wild type, it is often difficult to define this. Thus, maize is not known as a wild plant, and who is to say which human genotype is 'normal'? You will come across many systems in this Course and in reading other books, so we include a Table of accepted notations here.

A more comprehensive discussion of genetic notation is found in Units 3 and 4, Appendix 1.

Table 7   Accepted genetic notations for different groups of organisms

Group	Symbols used for the wild type	the mutant	Comments
fungi	*ad*$^+$	*ad* or *ad*$^-$	the $-$ superscript is usually left out
bacteria	*gal*$^+$	*gal*	three-letter code
viruses	*h*$^+$	*h*	
Drosophila	*vg*$^+$ or $+^{vg}$	*vg*	
mouse	*C*	*c*	
maize	*S*^{h}	*s*^{h}	
man	*Hb*^{A}	*Hb*^{S}	dominant mutations have capital letters
	HGPRT$^+$	*HGPRT*$^-$	
hypothetical loci	$\begin{cases} A \\ A^1 \end{cases}$	*a*	small letter as recessive
		A^2	different alleles but dominance relationship not shown.

2.9 Summary of Sections 2.0–2.8

Before going further with this Unit, let us summarize the position so far.

We have seen that DNA is the genetic material of living organisms, which in prokaryotes is present in the nucleoid(s) as 'naked', circular molecules (duplexes). In eukaryotes, however, it is organized in association with histones into complex, yet discrete, structures called chromosomes.

From genetic crosses, we can identify different characters, which are attributable to genes. We can recognize genes from these crosses, because they segregate according to a predictable pattern or algebra, which was originally postulated by Mendel. In parallel with these events, we have seen that the chromosomes of eukaryotic cells also segregate in a very predictable way at either mitosis or meiosis, depending on whether somatic cells are being replicated or the genetic material is being prepared for the formation of gametes.

We now turn our attention to the problem of relating gene segregation to chromosome segregation, and ask whether there is a direct correlation between chromosomal and genetic effects.

2.10 Sex chromosomes and their genes

Towards the turn of the century, Henking had observed a deeply staining chromatin element while studying meiosis in the sperm-forming cells of a hemipterous insect (a bug). Henking noticed that this element passed to one pole during anaphase I, so that at anaphase II, only half the sperm received it and the other half did not. Henking labelled this structure X, because he was uncertain whether it was a chromosome or not. During the next few years, other workers made similar observations with various grasshoppers and established the structure as chromosomal in nature. However, the important cytological discovery came in 1905 when Wilson observed that whereas male grasshoppers had only a single X-chromosome, the female always had two. Wilson proposed that there must be a causal relationship between chromosomes and sex. He argued that sperms that received an X-chromosome would on fertilization of an egg cell (all containing an X) produce a female insect, whereas sperm that did *not* carry an X-chromosome would be potentially male-determining. Although this mechanism could explain sex determination in some insects, Wilson and other workers later found that other insects, including the fruit fly *D. melanogaster* did not possess the odd X-chromosome. Instead the female cells carried three pairs of homologous chromosomes (autosomes) and a pair of homologous sex chromosomes (XX), but in the male the sex chromosomes were non-homologous (an unequal pair). One member of this pair resembled the X-chromosome found in the female cells; the other odd chromosome was called by Wilson the Y-chromosome. Other sex chromosome patterns have also been identified (Table 8).

sex chromosomes

Table 8   The sex chromosome differences between males and females in a variety of organisms

Organism	Male ♂	Female ♀
D. melanogaster	XY	XX
mammals	XY	XX
birds	XX	XY
butterflies and moths	XX	XY
grasshoppers	XO	XX
cockroaches	XO	XX
axolotl (Amphibia)	XX	XY
certain plants, e.g. white campion (*Silene alba*)	XY	XX

The sex that has homologous sex chromosomes (XX) is said to be *homogametic*; all the gametes it produces have an X-chromosome. In contrast, the sex that has an odd sex chromosome (XO) or non-homologous sex chromosomes (XY) is said to be *heterogametic*; half the gametes it produces determine the same sex, and the other half determine the opposite sex.

homogametic

heterogametic

ITQ 13 Is the male heterogametic or homogametic in:

(a) moths,

(b) mammals,

(c) cockroaches?

With this information to hand, we can now present further evidence that certain genes are associated with specific chromosomes (i.e. the sex chromosomes).

In 1910, T. H. Morgan by chance found a recessive mutant male fly with white eyes in a pedigree culture (wild-type) of *Drosophila melanogaster* (see *Life Cycles** and *HIST*). When this white-eyed male was mated with its red-eyed sisters, all the offspring produced in the F_1 generation were red-eyed. When males and females of the F_1 were interbred, the F_2 generation produced 2 459 red-eyed females, 1 011 red-eyed males and 782 white-eyed males. Morgan hypothesized that the white-eyed character was carried on the X-chromosome, but not on the Y-chromosome.

So how did Morgan arrive at this hypothesis? We can explain it best by a series of diagrams (Fig. 26) in which w = allele for white eye character and $+^w$ = allele for red eye.

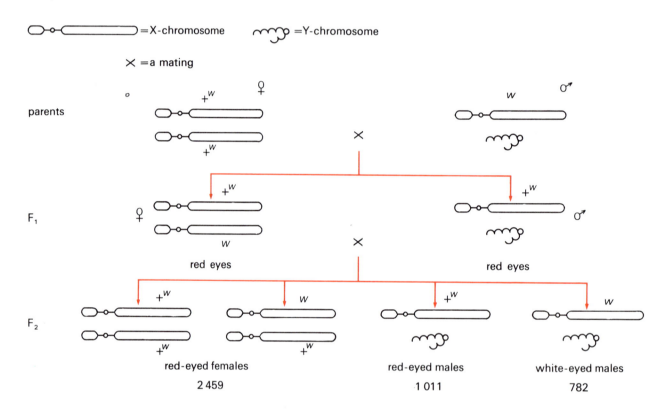

Figure 26

If, as Morgan hypothesized, the Y-chromosome of *Drosophila* is inert genetically (this was subsequently shown to be true), then half of the F_2 progeny will be female and half male. All the females will have red eyes, but only half of the males will have red eyes and the other half white.

The numbers obtained by Morgan almost fit the scheme above, although there are fewer males than expected, particularly white-eyed males. So are we justified in saying that Morgan's hypothesis is correct?

There is a means of checking whether the figures could come about by chance. If we apply the χ^2 test (see *STATS* and our working of the test earlier in this Unit, p. 84), we find that these figures could occur by chance less than 1 in 100 times. Either the hypothesis is open to doubt or else another factor needs to be considered. Fortunately, the reduction in the expected number of males can be accounted for by the lower viability of the males in this particular stock, especially the white-eyed males.

* The Open University (1976) S299 LC *Life Cycles*, The Open University Press. This folder, containing details of organisms mentioned in the Course, is part of the supplementary material for the Course.

However, other genetic matings with the same type of flies could be made to support Morgan's hypothesis.

ITQ 14 Can you suggest other matings that would verify the hypothesis?

During the course of maintaining stocks of *Drosophila*, Morgan found other recessive mutations in addition to the white eye: notably, miniature wing (*m*) and yellow body colour (*y*), which were also carried on the X-chromosome and behaved in a similar way to *w*. Any gene that is shown to be carried on a sex-chromosome is said to be *sex-linked*, and we shall be discussing sex-linkage and *autosomal** linkage in the next two Units (Units 3 and 4).

sex-linked
autosome

Now that we have discussed the segregation of genes from genetic crosses and seen that, in the sex chromosomes at least, there is a direct correlation between chromosome segregation and genetic effects, we can postulate the important genetic consequences of meiosis.

1   Meiosis serves to reduce the chromosome complement from the diploid to the haploid number. In doing so, it *segregates allelic differences*.

2   Meiosis allows the random segregation of non-homologous centromeres into different nuclei. This has the effect of *recombining non-allelic differences on non-homologous chromosomes*.

3   During meiosis, chiasma-formation normally occurs. We shall see in Unit 3 that chiasmata play an important mechanical role in regulating segregation. The exchange of segments between homologous chromosomes that occurs in chiasma-formation results, of course, in the *recombination of non-allelic differences between homologous chromosomes*.

4   When we bear in mind, too, that normally these genetic features are occurring in the formation of both egg and sperm and that many gametes with different genotypes are found, with further randomization entering into the determination of the genotype of the zygote through the choice of mate and the chance of which sperm fertilizes any given egg, we can see that the genetic 'shuffling and mixing' brought about by meiosis and fertilization are considerable.

All these points will be taken up again in later Units, but *to test your understanding of them at this stage, answer ITQ 15*.

ITQ 15 Figure 27 shows (a) female and (b) male germ-line cells at the beginning of meiosis. Three pairs of homologous chromosomes are present; each homologous pair is seen to carry alleles of specific genes (*A* or *a*; *B* or *b*; *C* or *c*).

(a) Write down the possible patterns of segregating genes in both female and male.

(b) How many different combinations are possible from a fusion between egg and sperm?

(c) If both male and female lines were heterozygous for all the genes shown above, how many different combinations would be possible from a fusion between egg and sperm?

(a) female germ-line   (b) male germ-line

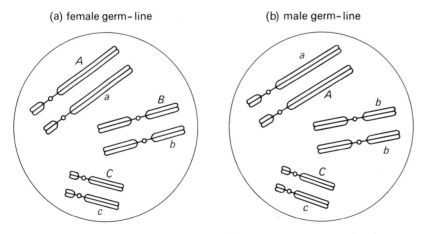

Figure 27

* Autosomes are all the chromosomes other than the sex chromosomes.

2.11 Gene mutation

Heritable differences between organisms arise from changes in genes or recombinations of genes. These genes are recognized and identified by carrying out genetic crosses. For centuries plant and animal breeders have recognized differences in phenotype in so far as they have identified in plants and animals desirable characteristics and selectively propagated or bred from them to produce, for example, the strains of wheat or the Hereford cattle that we know today. For the most part they had no conception of genes. The appearance of a change in the phenotype of an organism between one generation and the next, associated with a change in its genetic composition or genotype, provides the raw material of variation and hence is the basis of the study of genetics. The process giving rise to changes in genes is called *mutation*. It is basically a disruption in the constancy of the genetic makeup of an organism. It results from a change at the nucleic-acid level, irrespective of whether single or double strands of DNA or RNA are involved (as in bacteria and viruses) or complex structures such as the chromosomes of higher organisms. Pressing the analysis further, some sort of change in the genetic message encoded in the DNA or RNA is the root cause of all mutation. This is reflected directly or indirectly in either the presence or absence of a specific protein; or as a change in one or more of the proteins in an organism; or in similar effects on the RNAs involved in the synthesis of proteins. The molecular aspects of mutation will be dealt with at greater length in Unit 6. For the moment we shall concern ourselves with gaining some idea of the diversity of the products of mutation (*mutants*) that are known, how they are detected, at what sorts of frequencies they occur, and how that frequency can be increased.

mutation

mutants

2.11.1 The diversity of mutants and their detection

The following instances of mutants were mentioned in S100, Unit 19.

1   *E. coli* bacteria that are resistant to destruction (lysis) by the bacterial virus (bacteriophage) T4. The shortened term is 'T4 phage resistance'.

2   Sickle-cell anaemia in man.

3   Grain colour and shape in the maize plant.

4   Wing size and shape in the fruit fly.

5   Resistance of bacteria to the antibiotics, penicillin, tetracycline and streptomycin.

6   Achondroplastic dwarfism in man.

7   Seed size in the broad-bean.

8   Dark and light colouring in mice.

9   Resistance of insects to the insecticide DDT.

10   Rats resistant to the effects of the rat poison 'Warfarin'.

11   Haemophilia in man.

12   Moulds able to use the plastic PVC as a source of food.

13   Speed of growth and rate of food conversion in pigs and chickens.

14   Resistance to myxomatosis virus in rabbits and changes in the virulence of the virus.

These 14 instances of mutants provide us with a good cross-section of mutant types. They *could* be grouped in a number of ways. For ease of discussion let us consider them as being (a) 'advantageous' or (b) 'disadvantageous' to the organisms concerned, or (for want of a better term at present) (c) neither advantageous nor disadvantageous.

> QUESTION   On the basis of differences and similarities among these mutants, see whether you can arrange them into the three groups outlined above.
>
> ANSWER   (a) *Advantageous*—T4 phage resistance; penicillin, tetracycline and streptomycin resistance; DDT resistance; PVC utilization; Warfarin resistance; speed of growth and rate of food conversion and myxomatosis resistance.

(b) *Disadvantageous*—sickle-cell anaemia; achondroplastic dwarfism and haemophilia.

(c) *Neither*—grain colour and shape; wing size and shape; seed size and dark and light colouring.

It would be interesting to know how many mutants you assigned to group (c)—if any! Decisions about the value of a particular characteristic to an organism can be difficult to make because 'advantageous' and 'disadvantageous' will depend on the environmental conditions. With this in mind, let us take a closer look at one of the so-called disadvantageous mutants.

QUESTION Is possession of the sickle-cell anaemia allele (Hb^S) in man *always* disadvantageous? (You may need a quick look at Unit 1 to help you to answer this.)

ANSWER No. In parts of the world where malaria is prevalent, individuals who are Hb^AHb^S heterozygotes are more resistant to malaria than those homozygotes not carrying the sickle-cell allele (Hb^AHb^A).

This example brings out the important point that putting mutants into defined groups has its hazards. To repeat what was said on p. 88, they are grouped only for ease of discussion! Before continuing the discussion of these groups we need briefly to consider the significance of haploidy or diploidy in relation to the expression of mutations.

Organisms such as viruses, bacteria, and moulds possess single sets of genes—they are haploid. For the most part, with the exception of viruses, they reproduce asexually by division to produce clones, that is, large numbers of identical organisms. A mutation in any individual of a clone can be expressed immediately and in all of its descendants. By contrast, all the cells except the gametes of man, maize, fruit flies, mice and other higher organisms possess two sets of genetic determinants (genes)—they are diploid (see *Life Cycles*). The chance of both copies of a gene being mutated together is extremely rare. Therefore, a single mutation will be expressed in the resulting heterozygote only if it is *dominant*. The vast majority of mutations are not expressed in diploids because they are *recessive*. This is an important distinction, because immediate expression of a mutant characteristic makes the recognition and isolation of mutants considerably less complex in organisms that are haploid than in those that are diploid.

dominant
recessive

Group (a) Advantageous mutants

This group includes a sharply defined sub-group of mutants that can have, in certain situations, immediate and positive survival value.

ITQ 16 Which are the mutants with a 'survival kit' in their genes? Give reasons for your identification.

The rats and 'Warfarin' situation is slightly different because 'Warfarin'-treated food need not necessarily be the only food available, so that resistance to its effects may not be obligatory for survival. Increased speed of growth means that over a period of generations a mutant could outstrip the so-called wild-type organism and perhaps even eventually displace it in nature, becoming itself the 'wild type' at a later stage. But this is a very different situation from that demanding possession of the survival kit. (There is an interesting twist to this example—domestic animals that grow faster and convert food more efficiently or faster often get killed sooner to ensure supplies of meat!)

Mutants with a selective advantage (Unit 9) have been exploited widely, particularly by microbial geneticists in mapping and other genetic studies (Unit 3). Finally, mutants resistant to inhibitors have often been exploited in determining the mode of action of the inhibitors. This is certainly so for the antibiotics streptomycin and penicillin.

Group (b) Disadvantageous mutants

Such traits as sickle-cell anaemia and haemophilia in man can result in death. These are called *lethal* mutations.

lethal mutants

On the other hand, dwarfism or blindness, although very deleterious, do not necessarily result in death directly. Such traits are often called *semi-lethal*. Yet a third type are mutants that are lethal in one situation, but not in another. They are termed *conditional-lethal*; a good example is a mutant that confers lethality at one temperature, but survives at another. All three types of mutant occur widely and are much more common than those of Group (a)—we cannot go into the reasons for this now. Accepting this fact, we can now spend some time considering how some lethal and semi-lethal mutants are detected, how their time of occurrence is pin-pointed and how some conditional mutants are isolated.

semi-lethal mutants

conditional-lethal mutants

One way of detecting the occurrence and subsequent inheritance of lethal and semi-lethal mutations is to examine family trees (Unit 1) noting the presence or absence of the deleterious trait among offspring. Pedigree analysis is particularly applicable to humans. An important point concerning the effects of diploidy must be remembered—dominant mutations will be expressed immediately and in both homozygotes and heterozygotes, but the phenotypic expression of recessive mutations requires homozygosity, that is, the presence of the mutation in *both* sets of chromosomes.

You will see in Figure 28 a pedigree analysis over four generations for the semi-lethal dominant mutation in humans that results in cataract. This information was collected 60 years ago, but still provides us with a good example.

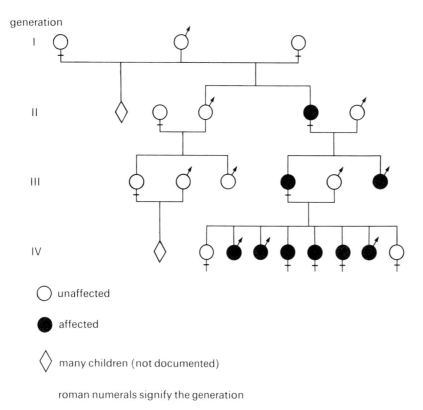

Figure 28 Pedigree for the inheritance of the semi-lethal mutation, cataract, in man—a dominant mutation.

ITQ 17 Where did the original mutation occur?

In Figure 29 (*opposite*) the pedigree is repeated, this time showing the postulated genotypes of individuals.

The original mutation is indicated by an X. It involved one chromosome of a homologous pair, one of which was derived from one parent and one from the other. For simplicity, only this pair of chromosomes is shown in the diagram.

Now look at Figure 30 (*opposite*) which shows the pedigree for the trait myoclonic epilepsy, another semi-lethal trait in man. The genetic defect was not known before generation I in this particular pedigree. This time you will see that the mutation is only revealed after marriages between cousins.

ITQ 18 Where did the original mutation occur, how was it inherited, and was it dominant or recessive?

generation

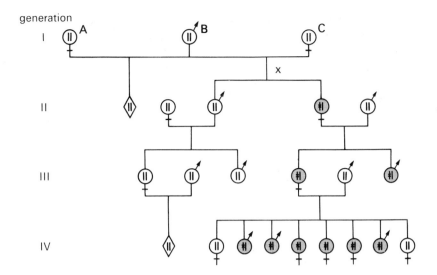

II ❙❙ homologous chromosomes
carrying normal gene

⬤ chromosome carrying
dominant gene

Figure 29

generation

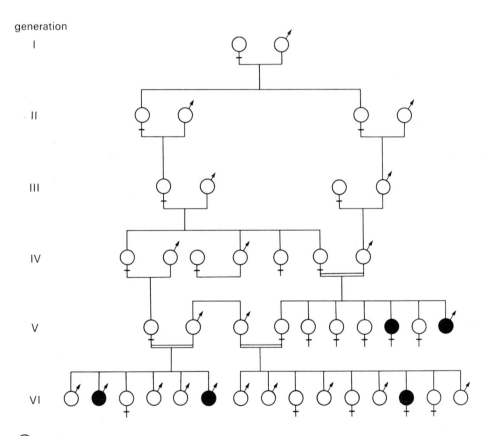

○ unaffected

⬤ affected

 marriages between cousins

Figure 30 Pedigree for the inheritance
of the semi-lethal mutation, myoclonic
epilepsy, in man.

Thus, from pedigree analysis we can deduce:

1 Whether a new mutation is responsible for the appearance of a novel phenotype.

2 Whether the mutation is dominant or recessive.

3 Approximately when in the pedigree the original mutation occurred.

91

Dominant lethal mutations result in the death of the organism and, unless this lethality is expressed after breeding has occurred, it will not be possible to study patterns of inheritance.

In the 1920s Muller designed a simple method for investigating recessive lethals in the X-chromosome of *D. melanogaster*. Subsequently, as you will see in Unit 8, it became possible to analyse lethals in the autosomes as well.

The sex of fruit flies is determined by their sex chromosome pattern, as seen in Section 2.10. You will recall that females have two X-chromosomes (the homogametic sex) and males an X-chromosome and a Y-chromosome (the heterogametic sex) in addition to three pairs of autosomes, as shown in Figure 31 below.

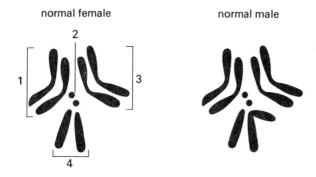

Figure 31 Autosomes and sex-chromosomes of *D. melanogaster* at mitotic metaphase.

At meiosis the gametes of the female each receive an X-chromosome and those of the male either an X-chromosome or a Y-chromosome. This means that half the progeny of matings should have two X-chromosomes and the other half an X-chromosome and a Y-chromosome, thus maintaining the 1:1 ratio of females to males. We shall focus attention on lethal mutations in the X-chromosomes. If a recessive lethal mutation occurs in the X-chromosome of the male gonads after X-irradiation or some other mutagenic treatment, it can be seen that all female offspring will inherit the X-chromosome carrying the lethal mutation (Fig. 32). They will, however, be heterozygous for the mutation, so that unless that mutation is dominant, the mutant phenotype will not be expressed and the sex ratio of 1 : 1 will be maintained.

sex-linked lethal mutation

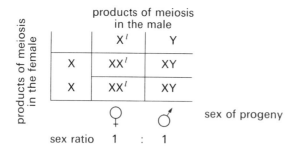

Figure 32 The effects of a recessive lethal mutation in a sex-chromosome of a male on the offspring of a mating with a normal female. The superscript *l* shows the occurence of a sex-linked lethal mutation.

QUESTION By a similar argument, what is the consequence of the occurrence of a recessive lethal mutation in one of the X-chromosomes of the female?

ANSWER

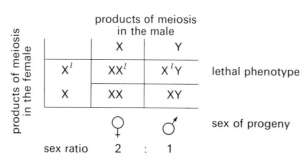

Figure 33

In this case it is obligatory that half the males of this mating acquire the X-chromosome carrying the lethal mutation. As each male has only one X-chromosome the lethality is expressed and these males die. The overall consequence is a reduction in the proportion of males, giving a ratio of females to males of 2:1.

In the fruit fly, the breeding experiments for identifying lethal mutants can be carried out under controlled laboratory conditions within the short period of 3 or 4 weeks. In contrast, the human pedigrees cover a century or more.

Let us now turn to a most important group of mutants whose lethality is conditional. In these, the phenotype of the organism is very much dependent upon the particular environment that the genotype encounters. Consider the following list of conditional lethal mutants (the years in which they were first reported are given in brackets).

1 Oat plants unable to grow in bright sunlight but able to grow in the shade. In these plants it was found that the chloroplasts were defective and as a result ineffective in strong light. (1922)

2 Some bacteria and moulds can grow on what is called a 'defined' agar medium containing simple inorganic salts and a source of carbon. Mutants requiring the addition of a wide range of different single amino acids, purines, pyrimidines or vitamins to such media can be readily isolated. They grow quite normally on a fully nutrient agar medium which might contain something like a meat or yeast extract. (1945)

3 Single bacterial virus (phage) particles, when surrounded by growing bacteria in nutrient agar medium, infect a single bacterium; their nucleic acid enters it and switches the bacterial metabolism over to the synthesis of phage components, which are then assembled into new phage particles and released by the destruction of the bacterial membranes and cell wall (lysis). As a consequence, several hundreds of these new particles are released for each bacterium lysed. These infect further bacteria and so the cycle of phage synthesis and bacterial lysis goes on until the bacteria stop growing. By this time the regions in the agar where the bacteria have escaped infection by phage are quite opaque—they form a dense lawn of many millions of bacteria. However, where single phage particles initiated successive rounds of infection and lysis, clear areas or plaques in the lawn of bacteria are formed. Mutants of the phage T4 form plaques on lawns of *E. coli* bacteria when incubated at 42 °C but not at 25 °C. (1964)

4 Certain mutants of the fruit fly are paralysed at 29 °C, but behave normally at 22 °C. This is due to a deficiency associated with the transport of sodium ions in the transmission of nervous impulses. (1974)

5 Most *Arabidopsis* seeds will grow in an agar medium containing essential nutrients. However, some mutant forms arise that require the addition of the vitamin thiamin for growth when the temperature reaches 27 °C, but they do not require it below a temperature of, say, 20 °C. (1955)

> QUESTION In the five examples, light, the supply of nutrients and temperature are the environmental conditions affecting lethality. Look through the examples again and identify the conditions that are lethal for each mutant.
>
> ANSWER (a) *light sensitivity*—oat plants; (b) *nutrient supply* (nutritional dependence)—bacteria and moulds; (c) *temperature sensitivity*—T4 phage and fruit fly; (d) *temperature sensitivity and nutritional dependence*—*Arabidopsis* seedlings.

temperature sensitivity
nutritional mutants

Do you see that (d) gives us another example of the artificiality of grouping mutations in living organisms? *Arabidopsis* mutants are *both* nutritionally dependent and sensitive to temperature; this makes groups based on any one conditional lethal characteristic unsuitable.

Temperature sensitivity is a mutation much used by geneticists, especially by microbial geneticists (Units 3 and 4, and 6). The lethal temperature can either be high (heat sensitivity) or low (cold sensitivity). Only examples of heat sensitivity occur in our list. Nutritional dependence (auxotrophy), as opposed to independence of the wild type (prototrophy), has been of great importance in studying both the metabolism and genetics of bacteria and fungi.

Procedures for the detection and isolation of conditional lethal mutants vary from one organism to another. The isolation of mutants of higher organisms such as plants and animals has almost exclusively resulted from careful observations of large stocks of wild-type organisms. In micro-organisms where very large numbers (in excess of 10^6) can easily be handled, a number of tricks involving the killing or removal of non-mutant organisms from populations have been perfected, so that finding the mutant becomes a simpler task. The isolation of auxotrophs of bacteria (Fig. 34) and fungi (Fig. 35) are good examples.

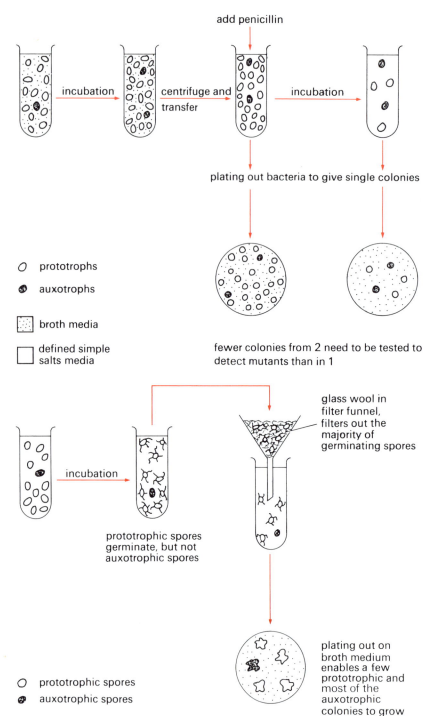

prototrophs

auxotrophs

broth media

defined simple salts media

fewer colonies from 2 need to be tested to detect mutants than in 1

Figure 34 An outline scheme for the isolation of auxotrophic mutants of bacteria using penicillin enrichment.

prototrophic spores

auxotrophic spores

Figure 35 An outline scheme for the auxotrophic mutants of the mould *N. crassa* using filtration enrichment.

Let us consider these examples in a little more detail.

Growing bacteria can be killed with penicillin because it interferes with cell-wall synthesis and makes bacteria so fragile that they burst. Prototrophic bacteria growing on a simple salts medium are killed, therefore, if penicillin is added. However, mutants unable to grow in the medium (auxotrophs) are unaffected by penicillin and survive. Consequently, in a bacterial culture containing many prototrophs and few auxotrophs, penicillin treatment increases the proportion of auxotrophs in the culture. The procedure is called penicillin enrichment. The auxotrophs are detected and isolated by allowing surviving bacteria to produce single colonies on a fully nutrient media that supports the growth of both prototrophs and auxotrophs. Each colony is then tested for the ability or inability to grow on a defined media.

In the bread mould *N. crassa*, prototrophic spores will germinate in a defined medium to produce small hyphae; auxotrophs will not. Germinated spores can be separated from ungerminated ones by means of filtration because the germinated spores have a relatively greater bulk. This results in an enrichment of auxotrophs similar to that achieved with bacteria.

Group (c) Neither advantageous nor disadvantageous

If you take our four examples in this group (grain colour and shape, wing size and shape, seed size and dark and light colouring in mammals), they are all 'visible' mutants. Other similar mutants are not directly visible, but can be detected. Three examples to be met in later Units and also included in Figure 36 are:

1   Plaque morphology mutants of phage-plaques formed by single phage particles on a particular bacterium. These have a characteristic morphology. Mutants with altered morphology can be easily detected and the mutant phage isolated from their plaques.

2   Sugar fermentation mutants of bacteria can be detected by adding a dye (eosin methylene blue) to agar medium. Normally, a single bacterium dividing by fission on the surface of nutrient agar will produce colonies containing about 10^6 or 10^7 bacteria. If the dye is added to the medium, together with different sugars such as lactose, galactose or mannose, it will react to any acidity caused by the fermentation of the sugars. Therefore, bacteria, which ferment sugars, will produce coloured colonies, whereas mutants, which do not, will give colourless colonies.

3   Colour of fungal colonies. This can reflect the colour of the mass of hyphae and/or the colour of the asexual spores or fruiting bodies formed above it.

Many visible mutants have been recognized and exploited in genetic studies of many different organisms. The list could be extended almost indefinitely by a visit to a cattle market, to a seed grower, to a pet shop or by taking a walk in a city park. In fact, if you did any of these things and looked carefully at similarities and differences between plants and animals you would be deriving for yourself a commentary on the detection and exploitation of visible mutations by man over the years.

Although these visible mutants are considered separately, in any given situation they are conditional. The size and shape of a wing or colouring could be of vital importance in, say, a predator/prey relationship in nature. So, the artificiality of grouping mutants in almost any way is underlined. One final point is that most new visible mutants in nature seem to be *deleterious* in some degree, although this may depend on the particular environment (e.g. melanism in moths; see Unit 9). In subsequent generations following the appearance of such a mutant in a population, the frequency of the mutant allele diminishes, the wild type being favoured. Also, new combinations of characteristics are notoriously difficult to retain in laboratory breeding populations.

2.11.2   The frequency of spontaneous mutations

The frequency of spontaneous mutation is very variable both among organisms and for different genes in any one organism. In higher organisms it can be expressed as the numbers of gametes tested for any one trait. This ensures some consistency between higher organisms and micro-organisms in the sense that the base line is the haploid state. To give you some idea of the frequencies of spontaneous mutation and the way in which they are expressed, look at the data below.

In a population of 51 380 wild-type grey-bodied (y^+) fruit flies, 6 yellow-bodied (y) individuals were detected. Thus, the frequency of mutation from $y^+ \rightarrow y$ was 6/51 380. This gives a percentage of 0.012, but often a negative index is used. Thus 0.012 per cent is the same as 1.2/10 000 or $1.2 \times 1/10\ 000$, which, expressed as a negative index, is 1.2×10^{-4}.

> **ITQ 19** Calculate the respective frequencies of mutation (as negative indices) from the following data.
>
> (a) T4 phage resistance in bacteria. From cultures of the bacterium *E. coli* containing on average 2.4×10^8 bacteria, the average number of T4 phage-resistant mutants per culture was 1.10.

(b) Sex-linked lethal mutation in the fruit fly. In an application of the test for sex-linked recessive lethal mutants (p. 92) to the Wooster strain of *D. melanogaster*, 8 recessive lethal X-chromosome mutations were detected among 1 266 females.

(c) Retinoblastoma in humans. A *dominant* mutation is responsible for this disease in man. It causes tumours in the retina of the eye during childhood and will cause death unless removed. Over a period of 9 years in Michigan, USA, 49 children out of 1 054 985 born to normal parents had the disease.

The frequencies of mutation in the examples of higher organisms which you have just handled (i.e. 6.3×10^{-3} and 2.3×10^{-5}) are higher (relative to the number of cell generations) than that for the bacterial example (i.e. 4.6×10^{-9}). Reliable data for fungi are difficult to obtain because of the predominantly hyphal nature of fungal growth and the widespread occurrence of the multinucleate condition. Nevertheless, the frequencies of mutation of fungi from wild type to auxotrophy, for example, are of the same order of magnitude as those for bacteria. On the whole, mutation frequencies in organisms such as animals and plants are higher than in micro-organisms, although the accuracy of estimating mutation rates in bacteria and viruses is better, because vast numbers of these organisms can be grown in a short time.

2.11.3 Comparison of the frequency of mutation among organisms

There are problems in making these sorts of comparisons, because the base lines or denominators for different organisms are often different. Mutation frequencies were expressed in Section 2.11.2 for diploids per gamete (for a dominant mutation) or per X-chromosome (for a sex-linked lethal mutation). The genetic determinants of haploid organisms such as bacteria and viruses are effectively the gametes of these organisms. Also, because each determinant is a single nucleic acid molecule, they are equivalent to single chromosomes. Thus, in this sense we are justified in expressing mutation frequency in similar terms for both higher organisms and micro-organisms. However, mutation frequencies are also expressed in terms of the genetic complement (or genome) of an organism. For bacteria there is only one chromosome, so the frequencies per gamete, per chromosome and per genome are the same, but for *D. melanogaster* an expression per genome requires the multiplication of the frequency per chromosome by the number of chromosomes, which is 4. Yet another way of expressing frequencies is per gene. The very approximate number of genes for man is calculated to be in the order of 100 000, for bacteria, 2 000 and for T4 phage, 56. Therefore, to express mutation frequencies in this way, we need to divide the frequencies per genome by the number of genes per organism.

To what extent are frequencies of mutation expressed per genome (total number of chromosomes) and per gene comparable for different organisms? This is really the question underlying the assumptions made in these calculations.

> QUESTION What assumptions are made about sizes of chromosomes and the susceptibility of genes to mutation when expressing mutation frequencies per chromosome or per gene?
>
> ANSWER The assumptions are that the sizes of chromosomes are the same, that the genomes are of similar size, and that all genes are equally susceptible to mutation. All of these assumptions can certainly be challenged!

Can a 'true rate' of mutation for any characteristic be determined? Or, putting it another way, is it possible to relate the frequency of mutation to time, that is, per generation? This can be achieved with bacteria that are growing at a constant rate in well-controlled conditions and where each cell division is effectively a generation. But the situation for higher organisms which contain many cells, of which only the gametes are concerned with the transmission of genetic characteristics from one generation to another, is very different and does not readily permit the calculation of true rates. Nevertheless 'rate' is used widely by geneticists, although strictly the figure is a fraction representing the number of mutants per given number of organisms in a population—a figure that in reality is itself a frequency!

Figure 36 (overleaf) Illustrated examples of visible mutations in a range of different organisms.

Figure 36 (overleaf) Illustrated examples of visible mutations in a range of different organisms.

(a) Plaques produced by two phages in a layer of *E. coli*. Plaques produced by the wild-type phase (r^+) are small and have a hazy peripheral zone referred to as a 'halo'. The halo results from lysis-inhibition by the wild-type phage. Rapid-lysis mutants (r) do not show lysis-inhibition and hence their colonies are large and have no halos. ($\times 1.2$)

(b) Colonies of *E. coli* on agar. Colonies at 24 hours have a uniform pale colour, characteristic of non-lactose fermenters. By 48 hours (shown here) lactose-fermenting mutants arise and grow above the original colony surface.

E. coli is a non-lactose-fermenting strain that produces lactose-fermenting variants; these appear as red papillae on the parent colonies. Papillae are largest in the heavily sown area of growth because as the non-lactose nutrients are exhausted, so the growth of the lactose-metabolizing variant is favoured. ($\times 2.8$)

(c) Growth of wild-type *N. crassa* (*left*) and one of its flat-colony mutants. The two strains were inoculated into separate tubes at the same time and incubated together. The wild type produces an abundant aerial mycelium which can be seen growing up the walls of its container. An aerial mycelium is lacking from the mutant progeny; its vegetative mycelium grows as a flat mat over the surface of the medium. The two variants have not been grown together on one plate because the wild type rapidly overgrows the mutant. ($\times 1$)

(d) Some of the variations in coat colour induced by mutations in mice. The normal wild-type coat colour is greyish-brown, but this is recessive to the dominant gene yellow colour ($A^y a$) shown here. Yellow, however, is expressed only in the heterozygous mouse, as it is lethal in the homozygous condition. The mutation (*da*) produces a darkening of the coat colour. The third mouse shown is the brown agouti type. The fourth mouse present is homozygous for the recessive mutant pallid coat (*pa*). As with many mutations, *pa* not only affects coat colour and pigmentation of the eyes (the eyes are pink), it also induces other effects, for example, defects of the inner ear.

(e) Mutations in aleurone (seed coat) colour and endosperm (internal food store) of maize seeds. The seeds are arranged on a cob. The normal, yellow, round-seeded cob is shown, but in the other cobs, there is a range of different mutations in aleurone colour, varying from greenish/blue to purple/pink, to yellow or almost white. There is also a mutation in some seeds from the normal plump endosperm to the recessive mutant shrunken endosperm.

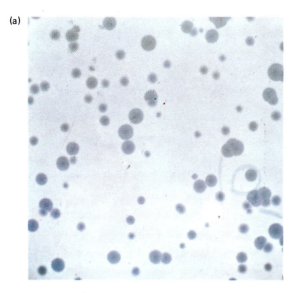

(a)

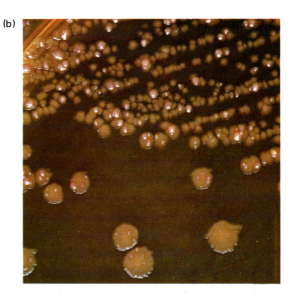

(b)

(c)

(d)

(e)

Table 9 shows the estimated rates of spontaneous mutations in a number of different organisms. Bear in mind that values in the higher organisms such as mouse and man are estimated from relatively few gene loci.

Table 9 Estimated rates of spontaneous mutation in different organisms

Organisms	Range of mutations expressed per gene locus per million cells or as gametes per generation
viruses	0.001–100
bacteria (*E. coli*)	0.001–10
maize	1–100
Drosophila	0.1–10
mouse	8–11
man	1–100

2.11.4 Environmental effects on the frequency of mutation

One way of illustrating an environmental effect, is to consider how increasing the X-irradiation can alter the frequency of sex-linked lethal mutations among the progeny of irradiated female *D. melanogaster*. In Figure 37 the percentage of these mutations is plotted against the X-irradiation dose in röntgen units (R)*, a convenient measure of the intensity of the dose.

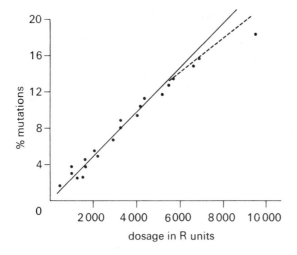

Figure 37 The frequency of sex-linked lethal mutations in the fruit fly following increasing doses of X-irradiation.

From the graph, we see that:

1 X-irradiation increases the frequency of mutation above the spontaneous level—it *induces* mutations.

2 The graph is linear up to 6 000 R. In other words, as the dose of irradiation increases during this phase the frequency of mutation increases in step with it. This indicates a one-to-one cause-and-effect relationship; one effective unit of irradiation brings about one mutation. At doses above 6 000 R, the curve in Figure 37 tends to depart from the linear (although the number of points on the graph are rather sparse at these high doses).

X-irradiation, then, can induce mutation and is an example of a mutagen. Ultraviolet light is also mutagenic; it does not have an ionizing effect but is absorbed to a great extent by the bases of DNA, which causes instability and induces abnormal bonding between them. In addition, many chemicals directly affect nucleic acid bases and are potent mutagens.

Mutagens can be:

1 Analogues of the bases (i.e. substances that closely resemble the bases in chemical structure and can, therefore, substitute for them).

mutagen

* One röntgen = 2.58×10^{-4} C kg^{-1}.

97

2   Agents that remove specific chemical groups such as amino groups from the bases and alter their nature *in situ*.

3   Substances that can insert themselves between bases and distort their spatial relationships.

4   Substances that make bases labile and permit substitutions.

· 5   Substances that interfere with the fidelity of nucleic acid replication.

Finally, some phages can induce mutations in their bacterial host cell. They do so, not by destroying the bacterium, but by inserting their nucleic acid into that of the host cell. In essence, therefore, all mutagens act directly or indirectly by interference with the genetic coding potential of an organism.

You will see from Figure 38 that the use of X-rays can increase the spontaneous frequency of mutation at least 100 times. Other mutagens are at least as effective. The use of mutagens, therefore, makes the isolation of mutants very much easier (pp. 97–8). To illustrate this, we may note that a mutation frequency of 2.3×10^{-5} for visible mutants of higher organisms can be elevated to $2.3 \times 10^{-3} - 2.3 \times 10^{-2}$ so that 1 in as few as 500 or even 50 survivors of mutagenesis will be a mutant. For conditional lethal mutations, such as auxotrophs of bacteria, a spontaneous frequency of 3×10^{-8} becomes 3×10^{-6} or 3×10^{-5}. It means that after penicillin enrichment (p. 94), there could be as many as 1 mutant in 300 or 1 mutant in 30 colonies of surviving bacteria.

There are mutagens that are even more potent than the examples we have considered. A mutation frequency of 1×10^{-2} among bacteria surviving mutagenesis is possible with some mutagens that interfere with the replication of nucleic acids. Perhaps not surprisingly, many mutagens are also potent inducers of cancer-like growths (carcinogens) in higher organisms and require very careful handling.

An interesting practical use of mutagens comes from their exploitation in pest control, particularly in the control of insect pests. Heavy irradiation or chemical mutagenesis of males before they are allowed to mate with normal females results, after mating, in a high incidence of fertilized but inviable eggs. In fact, it is often possible to see gross chromosomal breakages and we shall be discussing such events again in Unit 5. The males, therefore, are effectively sterile because their sperm carries a large number of dominant lethal mutations. Application of this genetic method of control was successfully carried out on localized populations of screw-worm fly on the island of Curaçao in the Caribbean. The larvae of this insect cause abscesses in the flesh of living mammals and are a serious pest to domestic animals. Once techniques had been devised for growing the insects on a large scale *in vitro*, sterile males (irradiated) were released in large numbers into the natural population and, in Curaçao at least, this virtually eliminated the pest within relatively few generations. The method was also effective for this particular pest in south-eastern parts of the USA, but it has only been partially successful elsewhere.

2.12   Mutants with the same phenotype

In this Unit we first considered the nature of genetic material, then looked at the way it replicates both at the molecular level and the chromosomal level before turning to the recognition of genes by segregation and their variation through mutation. The frequency of mutation is high enough to permit the collection or isolation of many mutants of an organism with the same phenotype (particularly if mutagens are used). But are all mutants of the same phenotype genetically identical, that is, of the same genotype? This is not only a fundamental genetic question but also has great practical value as you will see in Units 3 and 4 and Unit 6. Consideration of this question also allows us to introduce the important concept of complementation.

2.12.1   The complementation test

Consider the following situation in *D. melanogaster*. We have accumulated several mutants, all having a similar recessive phenotype, brown eye colour, which is an alteration in eye colour from the normal wild-type red. All our mutants are of independent origin, that is, they were found at different times and in different stocks, which previously had never shown that mutation. Each mutant can be fairly easily bred to form a double recessive homozygous stock, each stock carrying a specific mutation. Are all these mutant stocks genetically identical? We have said that their phenotypes (brown eye colour) are identical, but what of their genotype? A knowledge of the biochemistry of these pigments leads us to the discovery that all these mutants lack the activity of the enzyme xanthine dehydrogenase. So, even at the physiological level these mutants have identical phenotypes.

In following how we answer this question you will come to realize that 'identical' can have more than one meaning, depending on the criteria we adopt. The geneticist tackles the problem by crossing each stock in turn with every other and examining the F_1 phenotypes. Figure 38 shows the results of inter-crossing these stocks. Each box in the Figure shows the phenotype of the F_1 from a cross between the female of that column and the male of that row. Remember when you look at these results that you are looking at the phenotype of a diploid organism.

mutation in female parent

	1	2	3	4	5	
1	m	+	+	m	+	
2	+	m	m	+	m	
3	+	m	m	+	m	mutation in male parent
4	m	+	+	m	+	
5	+	m	m	+	m	

m = mutant (brown eye)

+ = wild type (red eye)

Figure 38 The phenotype of F_1 generations among brown-eyed mutants.

Notice that mating some stocks, for example, parent 2 with parent 1 (reciprocal crosses give similar results) produces a wild-type phenotype in the F_1. When this happens, we say that the two mutations *complement* each other. Whatever function is lacking in parent 2 is contributed by parent 1 and vice versa.

QUESTION Which of the other crosses produce complementation between mutant parents?

ANSWER 3 and 1; 5 and 1; 4 and 2; 4 and 3; 5 and 4.

So this rather basic test, in which a heterozygote is formed between two mutations, tells us that the mutations affect different functions and that, therefore, the mutations cannot be identical in function.

There is also an essential aspect of control in this test, namely, that each stock if bred to itself gives the brown eye-colour phenotype, as seen by the results in the bold diagonal line of boxes in Figure 38.

What we have done, therefore, is a *complementation test* on the different brown eye-colour mutants in *D. melanogaster*. As a result, we have been able to identify different mutations (genotypes) from what appeared to be essentially identical phenotypes. It should serve as an important reminder to you that you cannot presume *genetic* identity from *phenotypic* identity. Nevertheless, if we look at other mutations in heterozygous combination in Figure 38 (e.g. parent 3 with parent 2), we find that they do not give wild-type phenotypes in the F_1. As this result is the same as would be given by each stock if bred to itself, we can say that so far as this test goes, mutations 3 and 2 are identical, that is, they do *not complement* each other. Referring back to our 'explanation' of complementation, we have to say that within the limitations of the test, mutations 3 and 2 affect the same gene.

complementation test

The data of Figure 38 can be represented in an alternative way to give a *complementation map*. In making such a map we draw non-overlapping lines to represent mutations that do complement and *overlapping* lines to show mutations that do *not* complement.

complementation map

Thus, mutations 2 and 1 give ___2___ ___1___ , because they complement each other, whereas mutation 3 complements mutation 1 but not 2 so we can draw the map as shown:

1   2
_____ _____

3
_____

QUESTION Draw a complementation map to represent the other two mutations (i.e. 4 and 5 in Fig. 38).

ANSWER

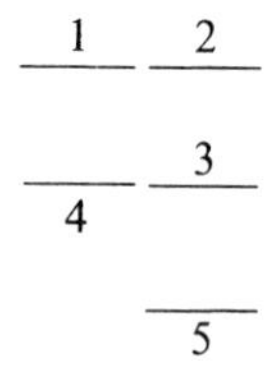

```
      1      2
   _____  _____

            ____3____
   _____
      4

             ____
              5
```

Mutant 4 is wild type in combination with 2, 3 and 5 so cannot overlap it, and vice versa for mutant 5. The complementation is believed to arise because the enzyme is made up of two different polypeptides, or two or more copies of a single polypeptide, and the mutations are believed to affect different parts of the enzyme molecule.

So far, we have carried out a complementation test with a diploid organism, the fruit fly, where we can obtain either homozygous or heterozygous individuals.

Can we carry out complementation tests with haploid organisms?

How can we obtain the 'heterozygote' necessary for a complementation test between two mutations introduced from different parents?

Let us take a case study from *N. crassa*, which is a haploid organism. In this organism, a number of histidine-requiring mutants are known which have arisen in different stocks. Like many other ascomycete fungi, *N. crassa* will form a *heterokaryon*, that is, hyphae in which nuclei from two different genotypes exist together in a common cytoplasm (see *Life Cycles*). If a heterokaryon is formed between two histidine-requiring mutants, and they complement, the heterokaryon will grow without histidine. Conversely, if they do not complement, then the heterokaryon will *not* grow without histidine.

heterokaryon

In a series of complementation tests with five different stocks showing a histidine requirement, the following results were obtained (Fig. 39).

mutation in 'female' parent

CD – 16	245	261	D – 566	1 438	
–	–	–	–	–	CD – 16
	–	+	+	+	245
		–	+	+	261
			–	–	D – 566
				–	1 438

Figure 39 N. crassa mutation in 'female' parent.

ITQ 20 Fill in the blank squares in Figure 39 and construct a complementation map from these data following the rules explained for the eye mutants of *D. melanogaster*.

This is a more complex situation than the rather simple example from *D. melanogaster*, but it is typical of many similar analyses with other mutants of microorganisms. It has significance in relation to the nature of the individual enzyme deficiences of different mutants.

Most of you doing this Course will be much less familiar with viruses such as phage than with animals and plants or even fungi. *As a final exercise in this Unit, recapitulate (if necessary) the way in which phage lyses bacteria in plaque formation (p. 93 and Life Cycles).*

On the same page (p. 93) reference is made to conditional lethal mutants of phage T4 which are sensitive to temperature (*ts*); they make plaques on *E. coli* bacteria at 25 °C but not at 42 °C. Many different T4 *ts* mutants have been isolated and so the 'same phenotype' situation is identical to that with the eye-colour mutants of *D. melanogaster* and the histidine-requiring mutants of *N. crassa*.

Given that a simultaneous mixed infection of individual *E. coli* bacteria with two different T4 *ts* mutants can be achieved, let us see how we can test complementation between *ts* mutants to see if their genetic defects affect the same or different functions.

In a haploid organism such as T4, one cannot obtain a true heterozygote. Nevertheless, an equivalent condition can be achieved when the gene products of the two parental T4 phages are mixed within the common cytoplasm of the infected bacterium. Thus, to do the equivalent of a complementation test, bacteria are infected simultaneously with two *ts* mutants and incubated at 42 °C. The wild-type T4 phage lyses bacteria at 42 °C; the *ts* mutants do not. Thus, if lysis (plaque formation) occurs, the two mutants complement; if it does not, they do not complement.

So you can see that the complementation test can be applied to very different kinds of organisms. This test tells us that many mutants with the same phenotype do not necessarily have the same genotype and the complementation map is a way of representing these results. It is a useful test, but as we shall see in later Units, it does have its limitations because it deals with gene products and not the interaction between genes.

2.13 Summary of Unit 2

At the beginning of this Unit (p. 58) we posed a number of questions. In answering them, first, we presented briefly some evidence that pointed to nucleic acids (particularly DNA) as the chemical basis of heredity and then we looked at the location of DNA in cells. Next, we considered the similarities and differences between prokaryotes and eukaryotes, particularly with respect to the organization of DNA in these two types of cell. In prokaryotes we showed that DNA is linear and is organized as a 'naked', circular duplex, whereas in eukaryotes, DNA is complexly associated with specific proteins called histones, which results in structures called chromosomes.

Chromosomes are visible only during cell division, of which there are two types, mitosis and meiosis. We described the principal features of mitosis and meiosis and outlined their genetic significance. We stressed that the mechanical separation of chromosomes during cell division should be viewed in the context of the complete cell cycle, and pointed out that DNA synthesis was primarily a feature of interphase.

A key concept in this Unit is the way in which genes are identified—by segregation analysis after a genetic cross. The direct relationship between the segregation of genes and chromosomes emerged when we were considering sex determination and sex chromosomes. We can be confident that chromosomes are the vehicles of inheritance and that they carry the genes. With genes defined, we then turned our attention to changes in genes, or mutations, thus introducing the concept of alternative forms of a gene (alleles). Some description of mutants and their detection was followed by a quantitative exercise designed to give an appreciation of the frequencies at which mutations take place. We also saw how it was possible to increase the frequency of mutation artificially by using different mutagenic agents. Finally, we showed how the complementation test can be used to reveal that mutants of the same phenotype do not necessarily have the same genotype.

In formulating his laws of inheritance without the knowledge of chromosomes, Mendel assumed that factors (genes) always segregated independently from each other. Mendel was fortunate in his selection of characters from the garden pea; they did all segregate independently. It is evident from this Unit, however, that there are more genes than there are chromosomes; you have encountered three different genes, white eye (w), miniature wing (m) and yellow body-colour (y) in our discussion of genes associated with the X-chromosome in *D. melanogaster*. These genes are said to be sex-linked. The concepts of linkage, independent assortment of genes and recombination, form the basis of Units 3 and 4.

Self-assessment questions

Section 2.3

SAQ 1 Label the following statements as *true* or *false*.

1 Bacteria and blue-green algae are referred to as prokaryotes, whereas other living organisms are referred to as eukaryotes.

2 Eukaryotes can be distinguished from prokaryotes by the differences in size of the ribosomal sub-units.

3 Eukaryotes contain DNA, but prokaryotes do not.

4 Prokaryotes are different from eukaryotes in that they lack membrane-bound organelles.

5 Eukaryotes are always diploid organisms, whereas prokaryotes are always haploid.

6 Organisms that have chromosomes enclosed within a nuclear membrane and distribute these to daughter cells by mitosis or meiosis are referred to as eukaryotes.

7 Eukaryotes can be both autotrophic* (photosynthesizers) and heterotrophic (non-photosynthesizers), whereas prokaryotes are only autotrophic.

Section 2.4

SAQ 2 Place the following mitotic features in the correct order as they occur during somatic cell division.

A The chromosomes are in their most contracted configuration having moved and gathered together into the equatorial plane of the spindle.

B The centromeres interact with the spindle apparatus.

C Division of the cytoplasm occurs.

D The chromosomes consist of long threads and are longitudinally double along their entire length, even at the centromeres.

E Sister chromatids move apart to opposite poles.

F The nuclear membrane breaks down and a spindle-shaped structure of microtubules is organized.

SAQ 3 A cytologist examines a number of slides of onion root-tip cells in various stages of mitosis. Of 1 000 cells counted, 692 cells were in prophase, 105 in metaphase, 35 in anaphase and 168 in telophase. What do you conclude from these observations?

Section 2.7

SAQ 4 A diploid organism has two pairs of chromosomes (four chromosomes in all). If one of its cells was observed when it was undergoing cell division under the oil-immersion objective ($\times$ 1 000) of a light microscope, how could you tell whether the cell was:

(a) in mitotic prophase or meiotic prophase I;

(b) in mitotic metaphase or meiotic metaphase I;

(c) in meiotic metaphase I or meiotic metaphase II;

(d) in pachytene or diplotene;

(e) in anaphase I or anaphase II?

---

* Autotrophic organisms are those that synthesize organic molecules from simple inorganic precursors (e.g. CO_2, H_2O, nitrates, etc) by trapping the energy from sunlight (i.e. all photosynthetic plants and some bacteria are autotrophs. Heterotrophs, on the other hand, are incapable of synthesizing organic molecules from simple inorganic precursors and rely instead on obtaining these large molecules (food) from autotrophs. Examples of heterotrophs are some bacteria, fungi and animals.

Section 2.8

SAQ 5 Red–green colour-blindness in man is caused by a sex-linked recessive gene c. A normal woman whose father was colour-blind marries a colour-blind man:

(a) What are the possible genotypes of the phenotypically normal woman?

(b) What are the chances that her first child will be a colour-blind boy? (You might need to check on probability in *STATS*, Section ST.2.)

(c) Of all the children from this marriage (sex unspecified), what proportion would you expect to be normal?

(d) Of all the girls produced, what percentage would you expect to be colour-blind?

Section 2.11

SAQ 6 A strain of wild-type bacteria can grow on or in a simple salts medium or a rich broth medium at both 25 °C and 40 °C. A liquid broth culture of this strain was treated with a chemical mutagen. Some of the mutants isolated from this culture showed the following growth responses on a simple salts medium containing either histidine or methionine at either 25 °C or 40 °C:

Table 10

Supplement to simple salts medium	None		Histidine		Methionine	
Temperature of incubation (°C)	25	40	25	40	25	40
1	+	−	+	+	+	−
2	−	−	+	+	−	−
3	−	−	−	−	−	−
4	−	+	−	+	+	+

+ indicates growth; − indicates no growth

All the mutants and the wild-type grew well on the rich broth medium at both 25 °C and 40 °C. From these data, what is the nature of each of the mutants 1–4?

SAQ 7 Arrange the following frequencies of mutation in ascending order of magnitude:

7.4×10^{-9}; 0.002 5 per cent; 6×10^{-4}; 0.23×10^{-5}; 1×10^{-8}; 0.001 per cent.

SAQ 8 Figure 40 shows the incidence of polydactyly (occurrence of extra fingers) in human beings.

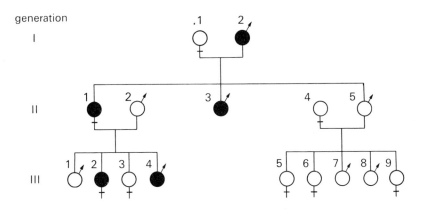

 normal

⬤ polydactylous

Figure 40

There are three genetic possibilities for its occurrence:

  (i) an autosomal recessive gene;

 (ii) an autosomal dominant gene;

(iii) a sex-linked recessive gene.

Which one of these three possibilities is the most likely and why?

SAQ 9 Label the following statements as *true* or *false* (if false say why).

1 Mutagens are substances that interfere with the fidelity of nucleic acid replication.

2 Most mutations occur while DNA is dividing; hence the mutation rate per generation is largely dependent on the number of cell divisions that occur in its germ-line.

3 Ultraviolet light is mutagenic.

4 A somatic mutation may cause physical death.

5 A mutation from the wild type to a dominant lethal in a single nerve cell would result in the death of a mature organism.

6 The frequency of mutation induced by X-irradiation increases more or less linearly with the total dose, but is not affected by the dose rate.

SAQ 10 H. J. Muller developed a system for detecting recessive sex-linked lethal mutations induced by X-irradiation in *D. melanogaster*.

(a) Assuming that spontaneous mutations are negligible in these experiments, calculate from the following data the expected number of mutants among 1 000 male progeny which received 6 000 R (röntgens):

54 mutations were detected among 723 male progeny receiving 2 500 R.
78 mutations were detected among 649 male progeny receiving 4 000 R.

(b) Taking a hypothetical example, let us say that in 5 different stocks of fruit flies that had each been given a 1 500 R dose of irradiation, a mutant phenotype with a blackish body-colour had arisen. After careful inbreeding to obtain the homozygous (recessive) form in each stock, the following complementation map was constructed.

stock 1	stock 2

stock 3	stock 4

	stock 5

Which of the stocks have the same genotype as well as the same phenotype?

Answers to entry test

1 True. DNAase is an enzyme that specifically breaks down (degrades) DNA.

2 False. Although RNA does contain the base uracil rather than thymine, RNA contains the sugar ribose whereas DNA contains the sugar deoxyribose.

3 True. Essentially bacteria are relatively simple unicellular organisms, although they may be found grouped in pairs or chains.

4 True. An enzyme is essentially a catalyst; it lowers the activation energy of a chemical reaction, but is not itself changed at the end of the reaction.

5 True. As far as we know all enzymes are proteins.

6 False. Ultra-centrifugation is used to separate various cell inclusions, but not a mixture of molecules.

7 True. A Geiger–Müller counter is a sensitive instrument for measuring radioactive emissions from radioactive isotopes of chemical elements.

8 True. Binary fission is a common form of reproduction in many unicellular organisms.

9 False. Amino acids are joined together by peptide links (*not* mRNA) to make proteins, although RNA determines the sequence of amino acids in a protein.

10   True. Viruses that attack bacteria are referred to as bacteriophage or simply phage.

The numbered structures in Figure 1 are:

1   cell membrane or plasma membrane;
2   nucleus;
3   mitochondrion;
4   rough endoplasmic reticulum (i.e. with ribosomes);
5   ribosomes;
6   cytoplasm;
7   membrane-bound vesicle;
8   smooth ER or Golgi apparatus;
9   nucleolus;
10   pores in nuclear membrane.

Answers to ITQs

ITQ 1   All that the data tell you is that the DNA content of diploid nuclei (in the chicken) is remarkably constant, whereas the protein content is extremely variable. Evidence is required that provides a direct relationship between DNA and the transmission of genetic information. At this stage in our argument, proteins, DNA and protein, or even some other large molecule(s) could be the hereditary material.

ITQ 2   There are several methods that you may have thought of:

1   The most obvious is to use a *specific* stain for DNA and observe microscopically where it is taken up in the cell.

2   Alternatively, you may have thought of extracting, purifying and then assaying DNA from cells; this method could give you a quantitative estimation of the amount of DNA in each cell, but it would tell you very little about its location. Better still, one could analyse DNA in subcellular fractions, for instance, of nuclei, mitochondria, microsomes, etc.

3   A third method, would be to take advantage of the specific chemical properties of DNA. Knowing that DNA undergoes synthesis during the life of the cell, it would be comparatively easy to follow the incorporation of a specific radioactively labelled precursor into DNA and hence to locate the position of DNA in the cell with a suitable monitoring device.

ITQ 3   The obvious differences are the absence of membrane-bound organelles (e.g. mitochondria) and endoplasmic reticulum in the bacterial cell, although it is possible that structures resembling ribosomes might be present in all three cell types.

ITQ 4   (*Objective 2*)   There is no obvious change in the appearance of the nucleoids of dividing *E. coli*. In contrast, the chromatin is much in evidence in many of the plant cells (those in mitosis or cell division) and chromosomes are quite distinct in the rat cell (also undergoing mitosis).

ITQ 5   Thymidine is a nucleotide specific to DNA. Uracil is a nucleotide specific to RNA.

ITQ 6   'Semi-conservative' replication means that the original double helices of DNA come apart during replication, but the half-helices or single DNA strands remain intact from generation to generation (i.e. they act as the template for the replication of 'new' DNA) and, therefore, the entire DNA component is 'semi-conserved' from generation to generation.

ITQ 7   The correct sequence is: 1B; 2G; 3C; 4H; 5A; 6F; 7D; 8E.

ITQ 8   (*Objective 4*)   *Half* the metaphase chromosomes in each cell would show one labelled and one unlabelled chromatid, and the other half would show no label in either chromatid.

ITQ 9 All the statements are compatible with Taylor's results. At present, geneticists are still uncertain about the number and arrangement of DNA molecules in the chromosomes of eukaryotes. Each chromosome may contain only one giant double helix of DNA or, alternatively, it might contain several double helices of DNA. (We shall be looking in more detail at chromosome fine structure in Unit 5.)

ITQ 10 The correct sequence is: 1A; 2D; 3G; 4C; 5F; 6H; 7B; 8E.

ITQ 11 Four types would be formed, each in equal numbers:
 (i) adenine-requiring and methionine-independent;
 (ii) adenine-requiring and methionine-requiring;
 (iii) adenine-independent and methionine-requiring;
 (iv) adenine-independent and methionine-independent.

ITQ 12 χ^2 is given by the *sum* of

$$\frac{(\text{deviation between expected and observed})^2}{\text{expected value}} \text{ for all four classes}$$

The expected value (equality of four classes) is $\frac{131}{4} = 32.75$.

$$\text{Thus} = \frac{(32.75{-}27)^2 + (32.75{-}30)^2 + (36{-}32.75)^2 + (38{-}32.75)^2}{32.75} = 2.5$$

From the table of χ^2 values with three degrees of freedom (see STATS, p. 55), $p \geqslant 0.50$; a deviation from equality as large or larger than that observed will occur by chance more than 50 times in 100. Therefore, these results are to be expected on the basis of two independently assorting genes.

ITQ 13 (a) homogametic, (b) heterogametic, (c) heterogametic.

It should be noted that some insects such as bees and wasps, hermaphrodite animals, fungi and many green plants, do *not* readily fall into this simple pattern of X, Y and O sex-determination.

ITQ 14 (a) The white-eyed male could have been mated with his (F_1) daughters. This cross ($X^wY \times XX^w$) should yield red-eyed females, red-eyed males, white-eyed females and white-eyed males in $1:1; 1:1$ proportions. (Morgan's figures were 129 red ♀, 132 red ♂, 88 white ♀ and 86 white ♂).

(b) Additionally, white-eyed females (X^wX^w) could then be *backcrossed* to the F_1 red-eyed males (XY), which should result in all the female offspring being red-eyed (XX^w) and all the males white-eyed (X^wY). This was also verified by Morgan.

ITQ 15 (a) Possible patterns of segregating genes are:

Table 11

Female	Male
ABC	abC
ABc	abc
AbC	AbC
Abc	Abc
aBC	
abC	
abc	
aBc	

(b) The possible number of combinations in a mating between female and male above is 32 (8×4), but the number of different combinations is 18 ($3 \times 2 \times 3$).

(c) If both female and male were heterozygous for all the gene loci above, then the possible number of different combinations at mating is 64 (8^2), but the number of different combinations is 27 ($3 \times 3 \times 3$).

ITQ 16 Bacteria and their viruses often occur in natural habitats such as streams, soils, or the guts of animals. Sensitivity to phage means destruction by lysis; resistance means survival. Also, bacteria that are resistant to antibiotics, or flies and mosquitoes resistant to DDT, or rabbits no longer susceptible to myxomatosis, are *all* survivors.

ITQ 17 In this pedigree, the spontaneous mutation occurred in one of the gametes of parents B or C in the first generation. From the second generation onwards, siblings who inherited the chromosome carrying the dominant mutation showed the mutant phenotype.

ITQ 18 Again this mutation probably occurred during the formation of the gametes in one of the two original parents (generation I) or their immediate ancestors. This resulted in the distribution of the mutation to the two daughters of marriage I. It was not expressed because it was recessive. After three, and again after four generations where marriages of cousins were involved, the mutant phenotype emerged among the homozygous siblings of generations V and VI. As in the answer to ITQ 17, only one pair of chromosomes is represented in the pedigree (Fig. 41).

generation

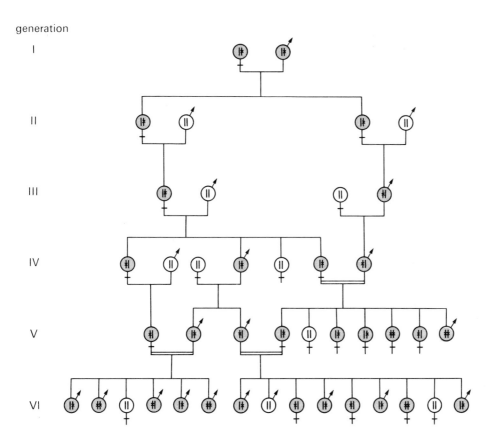

⊕ chromosome carrying recessive mutation which occurred
during gamete formation of generation I, or their near ancestors

Figure 41

When the mutant phenotype appears we can be certain about the chromosome situation for the individual concerned (the homozygous recessive) and also about that for the parents. In contrast, only a guess can be made about the siblings and also for some of the parents in the pedigree.

ITQ 19 There are an average of 1.1 mutants per 2.4×10^8 bacteria, so that the

$$\text{frequency} = \frac{1.1}{2.4 \times 10^8}$$

$$= 0.46 \times 10^{-8}$$

$$\text{or better } \underline{4.6 \times 10^{-9}}$$

(b) Frequency $= \dfrac{8}{1266}$

$$= \dfrac{8}{1.266} \times \dfrac{1}{10^3} \quad \text{or} \quad \dfrac{8}{1.266} \times 10^{-3}$$

$$= 6.3 \times 10^{-3}$$

(c) Frequency $= \dfrac{49}{10.55} \times \dfrac{1}{10^5} = \dfrac{49}{10.55} \times 10^{-5}$

As this is a dominant mutation and expressed in every heterozygote, it is necessary to multiply the frequency by $\frac{1}{2}$ because two haploid gametes are involved in the formation of every diploid zygote. The dominant trait, therefore, shows immediately in contrast to a recessive mutation.

Corrected frequency $= \dfrac{49}{10.55} \times \dfrac{1}{2} \times 10^{-5}$

$$= 2.3 \times 10^{-5}$$

ITQ 20 Complementation tests between histidine-requiring mutants of *N. crassa*.

mutation in 'female' parent

CD − 16   245   261   D − 566   1 438

−	−	−	−	−	CD − 16
−	−	+	+	+	245
−	+	−	+	+	261
−	+	+	−	−	D − 566
−	+	+	−	−	1 438

mutation in 'male' parent

− = no complementation
   (histidine-requiring)
+ = complementation (wild type)

Figure 42

The complementation map consistent with these data is:

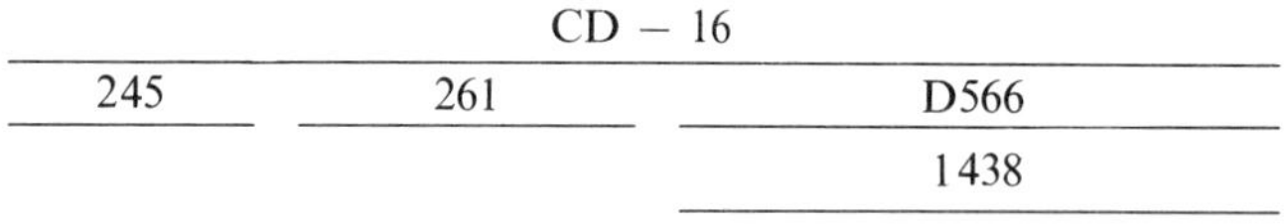

	CD − 16	
245	261	D566
		1438

Answers to SAQs

SAQ 1 (*Objective 2*)

1 True.

2 True.

3 False. (Clearly this answer is wrong!).

4 True.

5 False, although most prokaryotes are haploid, there are many eukaryotes (e.g. most fungi) that are also haploid.

6 True.

7 False. The statement about eukaryotes is true, but prokaryotes also can be autotrophic or heterotrophic.

SAQ 2 (*Objectives 3 and 5*) D, F, B, A, E, C.

SAQ 3 (*Objectives 3 and 5*) These observations give us an idea of the relative duration of each stage in mitosis—prophase is longest, then telophase, followed by metaphase, and anaphase is the shortest.

SAQ 4 (*Objective 3*)

(a) In mitotic prophase, four distinct chromosomes each consisting of two chromatids would be seen, whereas in meiotic prophase I, homologous chromosomes have synapsed and only two bivalents would be present.

(b) Mitotic metaphase would reveal four pairs of chromatids lined up independently on the spindle; meiotic metaphase would show two bivalents on the spindle.

(c) In meiotic metaphase I the chromosomes would be arranged as bivalents, but in meiotic metaphase II only univalents would be present.

(d) In pachytene (prophase I) homologous chromosomes would have paired (and would be shorter and thicker than in zygotene); each chromosome would still appear to be single. In contrast, diplotene homologues would be seen to be double-stranded and chiasmata would be present.

(e) In anaphase I, half the bivalents would be separating to opposite poles; in anaphase II the sister chromatids from each bivalent would be separating to opposite poles.

SAQ 5 (*Objectives 6, 7 and 8*)

(a) Either $+^c$ c

c or c

(b) One in four.

(c) One half.

(d) 50 per cent.

SAQ 6 (*Objectives 7 and 11*) All the mutants are auxotrophs. They are conditional lethal mutants which die in the absence of certain growth factors, but grow in their presence. Sometimes this auxotrophy also depends on temperature (another conditional lethality). The full descriptions of each mutant are:

1 Shows a requirement for histidine at 40 °C but not at 25 °C, that is, it is a *heat-sensitive* histidine-requiring mutant.

2 Shows a requirement for histidine at both temperatures, that is, it is independent of temperature.

3 Shows a requirement for a growth factor other than histidine or methionine (or both), and is independent of temperature.

4 Shows a requirement for methionine at 25 °C but not at 40 °C, that is, it is a *cold-sensitive* mutant.

SAQ 7 (*Objective 11*) 7.4×10^{-9}; 1×10^{-8}; 0.23×10^{-5}; 0.001 per cent; $0.002\,5$ per cent; 6×10^{-4}.

SAQ 8 (*Objectives 6, 7 and 8*) (ii) The pedigree data are most readily explained by an autosomal dominant gene.

The pedigree data are consistent with an autosomal recessive *provided* that the female in generation I is a carrier and male 1 in generation II is a carrier. The possibility that a sex-linked recessive gene is involved is *not* consistent with the data. It would be impossible to produce a polydactylous female in generation III, even if the mother in generation I were a carrier.

SAQ 9 (*Objectives 9 and 12*)

1 True.

2 True.

3 True.

4 True.

5 False. It is extremely unlikely that even a mutation to a dominant lethal in a *single, non-dividing* cell (i.e. a nerve cell) would result in the death of a mature organism.

6 True.

SAQ 10 (*Objectives 11 and 13*) (a) Mutations induced by irradiation are directly proportional to the dose given.

At 2 500 R the proportion of mutations is $\frac{54}{723}$ or 7.47 per cent.

At 4 000 R the proportion of mutations is $\frac{78}{649}$ or 12.02 per cent.

The difference = 4.55 per cent mutations for 1 500 R.

Among 1 000 progeny given 6 000 R, the expected number of mutants

$$= 1\,000 \times \frac{6\,000}{1\,500} \times 0.045\,5$$

$$= 182 \text{ mutants.}$$

(b) The following pairs of stocks complement each other: 1 and 2; 1 and 4; 1 and 5; 3 and 2; 3 and 4; 3 and 5. But the following pairs do not: 1 and 3; 2 and 4; 2 and 5; 4 and 5. Thus, stocks 1 and 3 have the same genotype as they do not complement, and stocks 2, 4 and 5 also have a common genotype for the same reason.

Bibliography and references

1 General reading

Goodenough, U. and Levine, R. P. (1974) *Genetics*, Holt, Rinehart & Winston.
(Chapters 1 and 2 and the first half of Chapter 5 are useful reading for this Unit.)

Lewis, K. R. and John, B. (1972) *The Matter of Mendelian Heredity*, 2nd edn, Longmans.
(Chapter 1 is especially useful.)

Watson, J. D. (1970) *Molecular Biology of the Gene*, 2nd edn, W. A. Benjamin.
(Chapters 3 and 9 are of special interest.)

White, M. J. D. (1973) *The Chromosomes*, 6th edn, Chapman & Hall.
(Chapters 1, 2, 3, 6 and 9 are recommended.)

Whitehouse, H. L. K. (1974) *Towards an Understanding of the Mechanism of Heredity*, 3rd edn, Edward Arnold.
(Chapters 11 and 12 are particularly useful reading for this Unit.)

2 References to S100

1 Unit 17, *The Genetic Code: Growth and Replication*
   Unit 19, *Evolution by Natural Selection*

2 Unit 17

3 Units 17 and 19

Acknowledgements

Grateful acknowledgement is made to the following sources for material used in this unit:

Figure 1 from S. Hurry, *The Microstructure of Cells*, John Murray; *Figure 5(a)* from M. Tribe *et al.*, *Electron Microscopy and Cell Structure*, Cambridge University Press; *Figures 6(a) and 7(a)* A. Ryter, Institut Pasteur; *Figure 6(b)* from M. C. Ledbetter and K. R. Porter, 'Introduction to the fine structure of plant cells' in A. L. Lehninger, *Biochemistry*, Springer-Verlag, 1970; *Figure 6(c)* from K. R. Porter and M. A. Bonneville, *Fine Structure of Cells and Tissues*, 3rd edn, 1968, Lea and Febiger; *Figure 7(b)* Courtesy of Dr Pickett-Heaps; *Figures 7(c) and 13* M. Anderson in W. A. Jensen and R. B. Park (eds), *Cell Ultrastructure*, Wadsworth Publishing Co. Inc., 1967; *Figure 8* from M. Meselson and F. W. Stahl in *Proceedings of the National Academy of Sciences*, USA, No. 44; *Figure 9* from J. Cairns in *Cold Spring Harbor Symposium* 1963; *Figures 12, 14, 21 and 23* after micrographs in Lewis and John, *The Matter of Mendelian Heredity*, 2nd edn. 1972, Longmans; *Figures 15 and 16* redrawn from J. H. Taylor, P. S. Wood and W. L. Hughes, in *Proceedings of the National Academy of Sciences*, USA, No. 43; *Figures 22 and 24* from M. J. D. White, *The Chromosomes*, 6th edn, Edward Arnold; *Figure 36(a–c)* R. J. Olds, Department of Pathology, University of Cambridge; *Figure 36(d)* M. E. Wallace, Department of Genetics, University of Cambridge; *Figure 36(e)* G. E. D. Tiley, Maize Unit, Wye College, Kent; *Figure 37* after W. P. Spencer and C. Stern, 'The relationship doses of X-rays and the percentage of X-chromosome lethals included in *Drosophila melanogaster*', *Genetics*, **33**, No. 43, 1948.